PEM Water Electrolysis

Hydrogen and Fuel Cells Primer Series

PEM Water Electrolysis

Volume 2

Dmitri Bessarabov

Pierre Millet

Series Editor

Bruno G. Pollet

ELSEVIER

ACADEMIC PRESS

An imprint of Elsevier

Academic Press is an imprint of Elsevier
125 London Wall, London EC2Y 5AS, United Kingdom
525 B Street, Suite 1650, San Diego, CA 92101, United States
50 Hampshire Street, 5th Floor, Cambridge, MA 02139, United States
The Boulevard, Langford Lane, Kidlington, Oxford OX5 1GB, United Kingdom

Notices
Knowledge and best practice in this field are constantly changing. As new research and experience broaden our
understanding, changes in research methods, professional practices, or medical treatment may become
necessary.

Practitioners and researchers must always rely on their own experience and knowledge in evaluating and using
any information, methods, compounds, or experiments described herein. In using such information or methods
they should be mindful of their own safety and the safety of others, including parties for whom they have a
professional responsibility.

To the fullest extent of the law, neither the Publisher nor the authors, contributors, or editors, assume any
liability for any injury and/or damage to persons or property as a matter of products liability, negligence or
otherwise, or from any use or operation of any methods, products, instructions, or ideas contained in the
material herein.

British Library Cataloguing-in-Publication Data
A catalogue record for this book is available from the British Library

Library of Congress Cataloging-in-Publication Data
A catalog record for this book is available from the Library of Congress

ISBN: 978-0-08-102830-8

For Information on all Academic Press publications
visit our website at https://www.elsevier.com/books-and-journals

Working together
to grow libraries in
developing countries

www.elsevier.com • www.bookaid.org

Publisher: Joe Hayton
Acquisition Editor: Raquel Zanol
Editorial Project Manager: Mariana L. Kuhl
Production Project Manager: Surya Narayanan Jayachandran
Cover Designer: Victoria Pearson

Typeset by MPS Limited, Chennai, India

CONTENTS

Dmitri Bessarabov, PhD, Director: DST National Center of Competence "HySA Infrastructure" at North-West University and CSIR, South Africa. Dr. Dmitri Bessarabov joined DST HySA Infrastructure Center of Competence at NWU and CSIR in 2010. He was recruited for a position from Canada as an internationally recognized scientist with academic and industrial decision-making experience in the area of membranes, hydrogen and electrocatalytic membrane systems for energy applications, and fuel cells. He received his PhD from Stellenbosch University, South Africa in membrane technology for gas separation. Dr. Dmitri Bessarabov is an ex-Aker Kvaerner Chemetics, ex-Ballard Power Systems, and ex-AFCC senior research staff member (all in Canada, British Columbia). Dmitri is also NRF (National Research Foundation)-rated scientist in SA. His expertise includes membranes, MEAs, and CCMs. He was leading CCM and MEA integration research team at AFCC for HyWAY 5 automotive program. His current responsibilities at HySA include: Leadership in the National Hydrogen and Fuel Cell Programme (HySA), Providing excellence in management, research and product development, HySA Infrastructure Business Plan development and implementation, Supply-chain and business development, engagement of industry into development activities. HySA Infrastructure 3-years rolling business plan includes: Establish technology platforms for PEM electrolyzer development, related characterization tools, electrochemical hydrogen compression (EHC) and separation, hydrogen storage and hydrogen production using renewables.

Pierre Millet, PhD, is an electrochemical engineer, Professor of physical-chemistry at the University of Paris-Sud, campus of Orsay. He graduated in 1986 from the French "Ecole Nationale Supérieure d'Electrochimie et d'Electrométallurgie de Grenoble" (ENSEEG), "Institut National Polytechnique de Grenoble" (INPG). He completed his PhD thesis on water electrolysis in 1989 at the French "Centre d'Etudes Nucléaires de Grenoble" (CEA-CENG). He is currently Director of the "Laboratory of Research and Innovation in

Electrochemistry for Energy applications" (http://www.icmmo.u-psud.fr/Labos/ERIEE/index_eng.php), at the French "Institute of Molecular Chemistry and Material Science" in Orsay. His research activities focus on water electrolysis, water photo-dissociation, carbon dioxide electro- and photo-reduction, hydrogen storage in hydride-forming materials, hydrogen compression and hydrogen permeation across metallic membranes. He is the author of 150 research papers and book chapters and has presented more than 150 oral communications at national and international conferences.

For decades proton-exchange membrane (PEM) water electrolysis has been used mainly for oxygen generation in anaerobic environments. Over the past two decades, however, it has been increasingly used for hydrogen generation in the industrial sector. The technology is also considered as a key one in the frame of the ongoing energy transition, for sustainable mobility applications and large-scale energy storage applications. For these reasons the technology is increasingly attracting attention. It is exciting to see how many research groups are now working on the subject, both experimentally and theoretically—in materials science and in electrochemical engineering. Many research articles, reviews, and books have already been published on the subject. Several national and international funding agencies are actively supporting R&D on the subject. Several companies have already put 1–10 MW-scale systems in the market and new announcements have recently been made for the next generation of 10–100 MW systems. Year after year, new materials become available, new designs are proposed, basic physical phenomena that underline the multiphysics features of the PEM reactor are better understood, performance and durability are increasing, systems are customized for new applications, capital expenditure is decreasing, and demonstration projects as well as market size are increasing. All these activities demonstrate that this topic is a vivid field of research.

In this context, the purpose of this book is to provide an overview on the subject, to highlight recent advances in the field and to discuss existing limitations that call for additional R&D. To attract the attention of readers the strategy behind the compiling of this book was to maintain a balance between Materials Science and Engineering data for hydrogen generation and PEM electrolysis, to provide information on key players in the PEM electrolysis sector, as well as highlight and review a few topics that either have not been receiving much attention or have recently received much attention, and furthermore, suggesting the direction of technology development for PEM electrolysis. These specific topics include, e.g., discussion on permeation and gas crossover measurements in PEM electrolyzers, defining efficiencies and key performance indicators for scale up. The authors believe that this book will provide a high-level overview of the state of the art for PEM water electrolysis technology, together with some new insights, not only pertaining to the basic properties of hydrogen, which is very useful, but also describing one of the methods of its production, i.e., PEM-based water electrolysis—all of which should make this book an attractive reference point for academics, scientists, engineers, and students.

ACKNOWLEDGMENTS

The authors of the book would like to acknowledge the assistance of Mr. Nicolaas Engelbrecht with preparation of Chapter 5, Selected Properties of Hydrogen, and Dr. Andries Krüger for his contribution in conducting gas crossover experiments, both of HySA Infrastructure at NWU. D. Bessarabov would like to acknowledge DST (Department of Science and Technology) of South Africa for support under HySA program. P. Millet would like to acknowledge the former French Energy Transition Institute "Paris-Saclay Efficacité Energétique (PS2E)" for providing support in the frame of the FlexiPEM research project on PEM water electrolysis.

CHAPTER 1

The PEM Water Electrolysis Plant

1.1 OVERVIEW OF THE PEM WATER ELECTROLYSIS SYSTEM

1.1.1 Laboratory Setup

The proton-exchange membrane (PEM) water electrolysis cell or stack, the core device used for the water dissociation reaction, needs to be placed in a specific environment for effective operation. For example, it is necessary to provide water of adequate purity and to supply that water to the cell/stack to feed the reaction but also to remove the excess heat produced by internal dissipation during operation at non-zero current density. It is also required to maintain the level of water purity during operation and to recycle the flow of electroosmotic water, to extract from the cell the reaction products (molecular hydrogen and oxygen) and to treat these gases in order to adjust their chemical composition to downstream requirements (in particular to remove oxygen traces and saturation water found in hydrogen at the outlet of the electrolysis module). All these different operations are achieved by a number of dedicated subunits that form together the so-called balance-of-plant (BoP).

Laboratory cells/stacks are mainly used for material and design qualification. What matters in such experiments is to adequately control operating temperature and pressure. Temperature is set to a constant value using a thermostat of appropriate capacity and operating pressure(s) are regulated using pressure controller. Fig. 1.1A shows the schematic diagram of a conventional laboratory setup used to measure the performances of laboratory PEM monocells or short stacks. The basic BoP functions are implemented. Two different water circulation loops are used to facilitate gas removal and isothermal operating conditions. Liquid−gas biphasic mixtures are separated by forced circulation in cylindrical chambers for cyclonic separation. Liquid water collected at the bottom of the separation chambers is recycled back to the cell/stack and the gaseous production (which is small for laboratory experiments) is usually vented. In some cases, oxygen traces are removed from hydrogen and hydrogen is dried whenever necessary.

PEM Water Electrolysis. DOI: https://doi.org/10.1016/B978-0-08-102830-8.00001-1

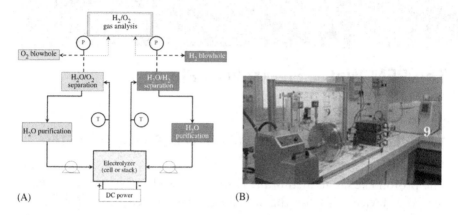

Figure 1.1 (A) Schematic diagram of the PEM water electrolysis laboratory setup. (B) Photograph of a typical laboratory setup for catalyst and design testing.

An online gas chromatograph can be used (Fig. 1.1B) to analyze the oxygen content of hydrogen, before and after purification. Thermocouples, pressure gas sensors, and mass-flow meters can be used for process monitoring. When durability tests are performed, online purification resins are also used to avoid cell contamination.

1.1.2 The PEM Water Electrolysis Plant

In a PEM water electrolysis (PEMWE) plant, the BoP is more complex and more sophisticated, due to the production unit and safety issues. Fig. 1.2 shows an overview of such plant (Smolinka et al., 2015). The different subunits are delineated by doted lines. The PEMWE unit (stack of electrolysis cells) is still at the heart of the overall process. DC electric power of appropriate specifications is supplied to the electrolysis unit by the power management unit that first uses an AC/AC transformer to adjust the voltage level of the main power source to the specific electrolysis plant requirements and then an AC/DC rectifier to produce the appropriate DC power source required by the electrolysis module. Purified water is provided by the water management unit that takes tap water and performs several purification steps required by the process. During operation, the water is pumped through the stack by the water circulation unit (WCU); online resins are used to maintain the required level of water purity and a heat exchanger is used to extract the excess heat released by the electrolysis unit. Hydrogen is released after conditioning (oxygen purification and water drying) in the gas conditioning unit (GCU). The electrolysis plant is placed under

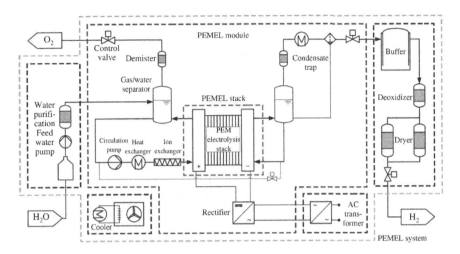

Figure 1.2 PEMWE process flowsheet (Smolinka et al., 2015).

Figure 1.3 Photograph of a containerized PEM water electrolysis module. Source: Courtesy AREVA H$_2$Gen.

the automated supervision of the process monitoring unit for continuous operation and monitoring.

Depending on its production capacity and the target application, the electrolysis plant can be either installed in a shipping container (the 40 ft container shown in Fig. 1.3 is the most frequently used; details are reported in Table 1.1) for easy transport and easy on-site implementation, or installed in a hall of an appropriate size to meet safety requirements and facilitate process installation, monitoring and management (Fig. 1.4).

In a PEM water electrolysis plant, there is usually no thermostat to adjust the temperature of circulating water. The plant is designed to operate at nominal current density and the heat produced under such

Table 1.1 System Packaging Details

Item	Unit	Reference Value
Container dimensions	ft	10–40
Container size	m	2.43 m wide, 2.59 m high, 12.2 m long
Container volume	m^3	67.7
Container weight	ton	3.75
Payload capacity	ton	27.6
Number of stack	per container	1–5

Figure 1.4 Photograph of a non-containerized PEM water electrolysis module. Source: Courtesy Siemens Co.

operating conditions is extracted and dissipated to the surroundings via a blower (strategies aimed at exploiting that heat for various applications in the surroundings of the plant have not been implemented yet). Usually, heat dissipation is the rate-determining step that dictates the maximum operating current density of the electrolysis unit. The system is designed for operation at nominal values. In Northern or Southern regions, during cold winters, the system can be operated above nominal conditions, because the extra heat can be easily extracted from the electrolysis module. Some typical packaging data of 40 ft shipping containers (ISO) are compiled in Table 1.1. Various container designs (tunnels with double doors at both ends, open-sided with doors down one complete side, double open-sided with doors down both complete sides, open-tops with a removable tarpaulin roof) can be used.

1.1.3 Comparison of PEMWE Systems

PEM water electrolyzers of different sizes (from the laboratory mono-cell to the multi-MW-scale plant) are nowadays commercially available from various suppliers. They are designed for several applications, are operated in quite different operating conditions, and can have somewhat different characteristics. Their comparison is therefore not a straightforward task. Several performance indicators (see Chapter 2, Key Performance Indicators) can be used for that purpose. The energy consumption per unit of hydrogen is of course one key performance indicator of interest for such comparison. At plant level, it is necessary to take into account all of the BoP including control and safety systems, power inverter, hydrogen purification, and any peripheries such as chillers, water purification units, and compressors. The different subsystems do not necessarily operate at all times. At high power density, the efficiency is dominated by the electrolysis stack while at low power density, the contribution of the BoP may become predominant. In transient power operating conditions (usually nonisothermal), it is possible to calculate the instantaneous energy consumption (and efficiency) but it is also necessary to integrate these values over a sufficient long period of time to obtain meaningful results. Except in the industry where water electrolyzers are used on an 8000 h•year^{-1} basis for the stationary production of hydrogen as a chemical, hydrogen energy applications (e.g., refueling station) do not operate continuously. In most cases, because of the limited number of fuel cell electric vehicles (FCEVs) on the road, 20% operation or even less are frequent. In outdoor conditions, when the electrolyzer is stopped, it is often required to turn the pumps on to heat the system and prevent the water from freezing. In some other cases of interest, the system may not produce but the pumps may be on to anticipate a quick start (e.g., for grid services). This is the reason why it is necessary to be careful when comparing the performances of PEM water electrolyzers. Such comparison of performance is possible only when the electrolyzers are operated under similar conditions.

1.2 THE WATER ELECTROLYSIS UNIT

1.2.1 The Conventional PEM Electrolysis Module
The general characteristics of a conventional PEMWE stack are compiled in Table 1.2.

1.2.2 From Mono-to Multistack Systems
Considering the investment costs needed to develop, test, qualify, and manufacture PEM water electrolysis stacks, technology manufacturers tend to develop a range of products that can cover the different market needs. Electrolysis stacks are not designed and constructed or custom-tailored according to customer needs. Of course, it is possible to adapt the number of cells in the stack but this in turn will have an impact on

Table 1.2 Main Characteristics of a Conventional PEMWE Cell or Stack

Cell Component	Element	Unit/Type	Reference Value
Cell/stack geometry	Geometry	Rectangular/circular	Circular
	Active area	cm^2	1–1500
	Cell per stack	Number	2–150
Polymer electrolyte	Supplier		
	Membrane type	PFSA Nafion	Nafion 115–117
	Equivalent weight	EW	900-1100
	Dry thickness	μm	150–180
	Reinforcement	Material	PTFE
	Electroosmosis drag	nH_2O/nH_3O^+	3–5
Catalyst	HER catalyst type	Supported/unsupported	C-supported Pt
	HER catalyst loading	$mg{\cdot}cm^{-2}$	1
	OER catalyst type	Supported/unsupported	Unsupported IrO_2
	OER catalyst loading	$mg{\cdot}cm^{-2}$	2
Catalyst layers	Manufacturing	Technology	Spray coating
	Composition anode		Catalyst + PFSA
	Composition cathode		Catalyst + PFSA
Membrane electrode assembly	Commercial availability	Yes/no	Yes
	Surface limitations	Yes/no	Yes
Other cell components	Bipolar plate	Type	mm thick Ti
	Flow field/disrupter	Type	Ti grid or flow field
	PTL anode	Type	Porous Ti
	PTL anode	Type	Porous Ti
	Sealant	Type	Elastomer
Operating conditions	Power	MW	10^{-6}–10
	Power density	$W{\cdot}cm^{-2}$	1.8–3.6
	Current density	$A{\cdot}cm^{-2}$	1–2.5
	Nominal cell voltage	V	1.8 at 1 $A{\cdot}cm^{-2}$
	Nominal temperature	$^\circ C$	50–70
	Maximal temperature	$^\circ C$	80
	Nominal O_2 pressure	bar	1–25
	Nominal H_2 pressure	bar	1–70
	Nominal ΔP	bar	10
Operation characteristics	H_2 cross-over	% of H_2 produced	<2%
	O_2 cross-over	% of O_2 produced	<1%
	H_2 dryer loss	% of H_2 produced	<1%
	H_2 in O_2	ppm or %	<1%
	O_2 in H_2	ppm or %	<1 ppm

EW, *equivalent weight;* HER, *hydrogen evolution reaction;* OER, *oxygen evolution reaction;* PFSA, *perfluorosulfonic acid;* PTFE, *polytetrafluoroethyle;* PTL, *porous transport layer.*

 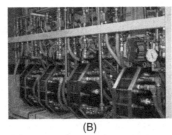 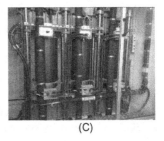

(A)   (B)   (C)

Figure 1.5 (A) PEM monostack. (B) PEM quadristack. (C) PEM tristack. Source: (A) and (B) Courtesy AREVA H₂Gen, Les Ulis, France. (C) Courtesy ITM Power Co., Sheffield, United Kingdom.

other BoP components and induce unnecessary extra costs. Taking advantage of PEM technology to operate over a large range of current density, multistack approaches are used to adapt the hydrogen production rate to customer needs. The implementation of several stacks instead of one monostack provides the necessary flexibility to the PEM water electrolysis unit (Fig. 1.5).

1.2.2.1 PEM Clusters

In this section, the term "cluster" is used to designate the multistack electrolysis unit and core BoP components such as power supply, water circulation circuits, and liquid–gas separators, as well as their interconnection. When it becomes necessary to interconnect several stacks to increase the production capacity, there are several options left to the engineering designer. The different units and subunits of the cluster can be factorized or not. Nonfactorized units can be connected either in series or in parallel. To choose one particular configuration, several criteria must be considered:

- *Maintenance strategy*: Is it necessary or not to stop one production unit while the others are still operating?
- *Subunits requirements*: Can units be operated separately or not? (e.g., what are the electrical characteristics of the main power supply and the power of circulation pumps?)
- *Operating parameters of PEM stacks*: To improve flexibility, is it better to stop one stack of a multistack cluster or is it better to turn all the stacks of the cluster down to a lower operating current density?
- *Economic criteria (Capex)*: What is the impact of the different options on cost?

- *Packaging considerations (wiring and tubing connection)*: Are there any spatial constraints that require a strong integration of the plant or not.

For maintenance purpose, it might be desirable to extract one PEM stack from the cluster without stopping production. In that case, configurations with individual power supply are preferable. When PEM stacks are wired in parallel, it is possible to extract one unit from the cluster by switching to a unit of equivalent impedance. When water is supplied in series, the mass flow must be sufficient to maintain enough water in the biphasic mixture. Fig. 1.6 shows an example of multistack configuration with individual BoP. This is the series connection of individual power supply units, individual water supply units, and individual gas separators. Small and not expensive pumps can be used; eventually, no pump is required when the cell/stacks operate at low current densities with natural gas lift phenomena. Such configuration is easier to manage and provides full flexibility, but of course there is a strong impact on Capex. This is well adapted to small stacks.

Fig. 1.7 shows an example of an alternative, more integrated, and multistack configuration. This is a series of connection of individual power supply, individual water supply but common gas separators. Such configuration still provides a large flexibility at a reduced Capex.

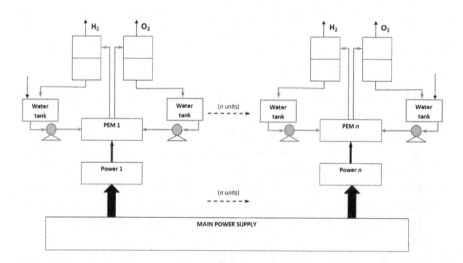

Figure 1.6 PEM water electrolysis cluster (configuration A).

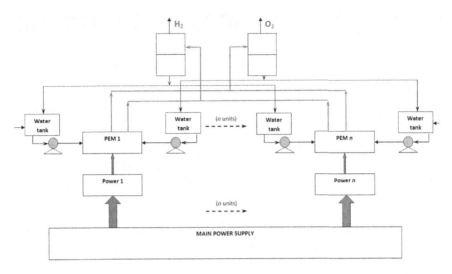

Figure 1.7 PEM water electrolysis cluster (configuration B).

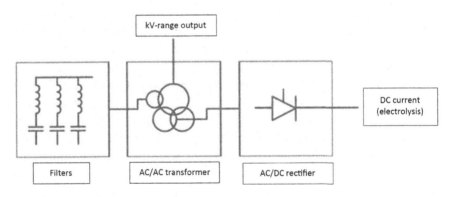

Figure 1.8 Schematic diagram of a DC station.

In many containerized commercial systems, integration should be maximum and all BoP units are factorized.

1.3 THE POWER SUPPLY UNIT

1.3.1 General Description

As shown in Fig. 1.8, *power units* contain an AC/AC transformer (to adjust the voltage level of the main power source to the specific plant requirements) and an AC/DC rectifier (to produce the necessary DC power source required by the electrolysis module). Filters are used to

Table 1.3 Some General Characteristics of AC/DC Rectifiers		
Item	Unit	Reference Value
Power rectifier	kW–MW	Process dependent
Rectifier efficiency	%	> 94
Rectifier cooling	Yes/no	Yes (air or water)
Active power (inlet)	kW	90–500
Reactive power (inlet)	kVA	Various
RMS voltage (inlet)	Vrms	Country dependent
Total harmonic distortion	%	< 10
Power factor	cos ϕ	0.96

treat harmonic distortion. The PEM water electrolysis plant does not require specific equipment compared to other similar electrolysis technologies.

Some specifications of the rectifier are compiled in Table 1.3. AC rectification is efficient but requires (especially for large systems) cooling to avoid detrimental heating. Active and reactive power have the usual meaning. The root mean square (RMS) voltage and current values are used to calculate the apparent power. Linear and time-invariant systems (LTIS), characterized by transfer functions, do not alter the frequency of passing signals. Non-LTIS systems induce distortions (power is transferred to the harmonics of the original frequencies). Total harmonic distortion measures the extent of that distortion. This is the ratio of the sum of the powers of all harmonic components to the power of the fundamental frequency. The power factor is the dimensionless ratio of the real to the apparent power flowing to the load (the electrical components that consume active electrical power). These different characteristics define the efficiency of the power unit which is considered as an auxiliary unit of the electrolysis process.

1.3.2 Examples

An example is provided by the PEM water electrolysis plant installed at the Energy Park in Mainz, Germany (http://www.energiepark-mainz.de/en/). Three DC stations (2 MW each) generate the DC current required for the electrolysis process (3500 A DC each). The electrical power required by each station is drawn directly from the medium voltage grid (20 kV, three phases) using GEAFOL cast-resin transformers. The current is inverted by industry-proven SINAMICS

Figure 1.9 Photograph of the AC/DC rectifier installed at the Energy Park in Mainz. Source: Courtesy Siemens Co.

inverter cabinets. Unwanted high grid interference is prevented by an integrated multiple-stage filter circuit system. Fig. 1.9 shows a photograph of the AC/DC rectifier installed at Energy Park in Mainz.

1.4 THE WATER PURIFICATION UNIT

1.4.1 General Description

The purpose of the water purification unit is to provide the necessary deionized water to the process. Most systems commercially available are using tap water (developed country standards) as process water. The use of seawater or any other source will require additional treatments to reach the necessary standards of purity/quality. Some general characteristics are compiled in Table 1.4. Water purity is a key prerequisite for long-term efficiency. Tap water is first treated and filtered to remove any bacteriological residue and then highly deionized. Purified water can be stored in an adequate storage tank. Air contamination is prevented by filters or inert gaseous skies.

During operation, online ion-exchange resins are placed on the water circulation loops to maintain conductivity within specifications.

Table 1.4 Main Characteristic of the Water Distribution/Management Unit		
Feed water minimum resistivity	MΩ•cm	1
UV treatment	Yes/no	Yes
Water purification efficiency	%	50
Use of purified water storage tank	Yes/no	Yes
Purified water tank size	m^3	System dependent
Pressure injection pump	Yes/no	Yes
Water circulation pump anode	Yes/no	Yes
Anode loop water conductivity	μS•cm^{-1}	0.1–1.0
Nominal anode water flow	λ (Eq. 4.3)	10–100 (system dependent)
Anode water flow regulation	Yes/no	Yes
Water circulation pump cathode	Yes/no	Manufacturer dependent
Cathode loop water resistivity	MΩ•cm	Same anode
Nominal cathode water flow	λ (Eq. 4.3)	N/A (or same anode)
Cathode water flow regulation	Yes/no	System dependent
Recycling of electroosmotic water flow	Yes/no	Yes

Table 1.5 Example of Resin Characteristics	
Maximum operating temperature	60°C (140°F)
Minimum bed depth	1000 mm (40 in)
Service flow rate	8–80 BV^a • h^{-1}
Maximum service velocity	120 m•h^{-1}
^{a}1 BV (bed volume) = 1 m^3 solution per m^3 resin.	

Examples (Table 1.5) are capsulated polystyrene cation and anion exchange resins, supplied in the H^+/OH^- form and containing a stoichiometric equivalent of the strong acid cation (the functional group can be a sulfonic acid) and the strong base anion (the functional group can be a trimethylammonium) resins. These resins should combine the properties of high capacity with excellent resistance to bead fracture from attrition and osmotic shock. The pressure drop is a function of operating temperature and service flow rate. Usually, thermal stability is limited and operating temperature to a maximum value of 60°C is specified.

1.4.2 Example

On small and containerized modules, the water purification unit (including tap water purification, storage, online purification, and circulation pumps) is enclosed in a dedicated cabinet. Fig. 1.10

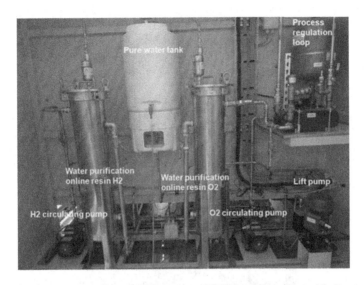

Figure 1.10 Water purification unit of a 130 kW containerized PEM water electrolysis module. Source: Courtesy AREVA H$_2$Gen.

Table 1.6 Main Characteristics of the Liquid–Gas Separation Unit		
Item	Unit	Reference Value
Water level monitoring	Yes/no	Yes
Online temperature regulation	Yes/no	No
Online cooler	Yes/no	Yes
Cyclonic separation	Yes/no	Yes

provides an example of such water distribution unit. In larger PEM plants, located in a production hall, the unit can be placed anywhere but preferably in the vicinity of the electrolysis modules to avoid the transfer of water over too long distances. In some cases, the electro-osmotic water flow collected at the cathode during electrolysis is sent back to the water storage tank instead of the anodic water circulation loop. This requires specific pressure management procedures but avoir risky interconnection of anodic and cathodic water loops.

1.5 THE LIQUID–GAS SEPARATION UNIT

1.5.1 General Description

Some general characteristics of the liquid–gas separation unit are compiled in Table 1.6.

Figure 1.11 Water circulation circuitry of a laboratory 25 kW PEM water electrolyzer.

1.5.2 Example

The WCU (Figs. 1.11 and 1.12) contains the water circulation loops placed at the anode and cathode of the electrolysis unit. Some technology suppliers use only one pump at the anode. Others use one on each side. Some systems operate at equi-pressure. Others operate under a difference of pressure. A purification unit is placed on the anodic loop as well as a heat exchanger to cool down circulating water and set nominal operating temperature. The electroosmosis flow of water collected on the cathode side is transferred back to the anode at regular time intervals or sent back to water treatment unit.

1.6 THE GAS TREATMENT UNIT

1.6.1 General Description

Gas (H_2) purification is usually included in the PEM water electrolysis plant. The level of gas purity is a function of the application and customer specifications. Main impurities are oxygen and water. Oxygen

Figure 1.12 Liquid-gas separators of a 2 MW PEM water electrolyzer. Source: Courtesy Siemens Co.

Table 1.7 Some Characteristics of a Chiller/TSA Gas Treatment Unit		
Item	Unit	Reference Value
Deoxo	kW	4
Chiller input/output temperature	°C input	60−70
	% input	Saturation
	°C output	3−5
	% output	Saturation
TSA power	$kW \cdot (10\ Nm^3_{H_2} \cdot h)^{-1}$	1
H_2 consumption for TSA regeneration	%	< 10
TSA input water content	%	Saturation at 5−7°C
TSA input O_2 content	ppmv	< 500
TSA output water content	%	< 5
TSA output O_2 content	ppmv	< 5
TSA output N_2 content	ppmv	< 2
Equivalent dew point	°C	−70

can be eliminated by passive catalytic recombination. Water can be eliminated by using a chiller of appropriate capacity and/or by temperature swing adsorption (TSA). Some general characteristics of the gas treatment unit are compiled in Table 1.7.

1.6.2 Example

The gas treatment unit usually contains a deoxo and a gas drying subunits. At the anode, the GCU (Figs. 1.13 and 1.14) contains only a demister (to remove and recycle saturation water vapor) and a control

Figure 1.13 Gas purification unit of a 300 kW containerized PEM water electrolyzer. Source: Courtesy AREVA H₂Gen.

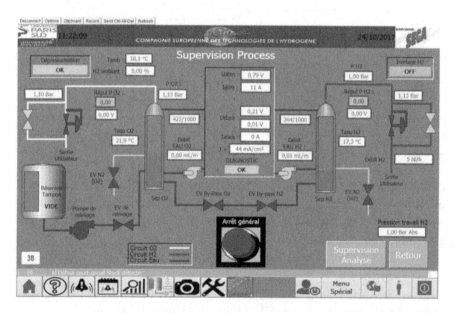

Figure 1.14 Process overview of a PEM water electrolysis laboratory unit.

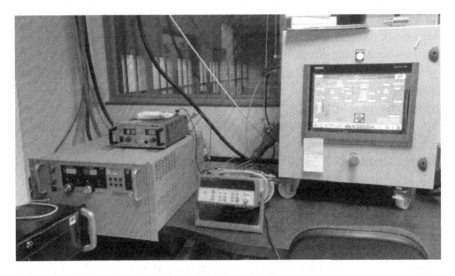

Figure 1.15 Laboratory process monitoring unit and DC power supply.

valve to adjust O_2 exhaust pressure. At the cathode, the water-saturated hydrogen is first cooled down to remove gas saturating water. Then oxygen traces are removed by catalytic recombination and resulting water is finally dried out.

1.7 THE PROCESS MONITORING UNIT

The process monitoring unit (Fig. 1.15) is used for the supervision of the process. It is usually designed for automated operation and in some cases for remote monitoring and/or operation. All operating phases (initiation, ramp-up, operation, shut-down, emergency actions) are automated. The process can be driven and supervised remotely, in some cases from overseas. Operating data can be transferred anywhere in the world via the Internet. Some technology suppliers may also offer manual operating conditions.

1.8 OVERVIEW OF SEVERAL TECHNOLOGY SUPPLIERS

The performances of several commercial PEM water electrolyzers are compiled in Table 1.8.

1.8.1 Siemens

Siemens (Erlangen, Germany, https://www.siemens.com/global/en/home. html) is developing large PEMwater electrolysis plants for energy management. These developments rely on a range of PEM stack technology called Silyzer. Details on the Silyzer technology developments can be found at the following weblink (https://www.siemens.com/silyzer). Fig. 1.16 shows an overview of the Silyzer 200 (1.25 MW) baseline stack technology commercially available and Table 1.9 provides some technical characteristics. The Silyzer 200 skid contains 5 stacks of 50 cells each. The active surface area is $1,400\ \mathrm{cm}^2$. The nominal operating current density is $1.7\ \mathrm{A \bullet cm}^{-2}$ and the peak current density is $2.5\ \mathrm{A \bullet cm}^{-2}$ (50% more). The H_2 delivery rate is $49.7\ \mathrm{Nm}^3{}_{H2} \bullet \mathrm{h}^{-1} = 4.4\ \mathrm{kg}_{H2} \bullet \mathrm{h}^{-1}$. The next generation of PEM electrolyzer, Silyzer 300, is currently under development. The objective of the company is to deploy on the market systems that can address the $10-100\ \mathrm{MW}$ power range, with an overall efficiency of 80% and a target price of €$600-800\ \mathrm{kW}^{-1}$.

Table 1.8 Some Manufacturers of PEM Water Electrolyzers and System Characteristics				
Manufacturer	Country of Origin	Capacity Range $(\mathrm{kg}_{H2} \bullet \mathrm{h}^{-1})$	Pressure (bar)	Energy Consumption Range $(\mathrm{kWh} \bullet \mathrm{kg}_{H2}{}^{-1})$
Siemens	Germany	0−60	1−35	45−65
Proton OnSite	United States	0−21	1−30	47−55
Hydrogenics	Canada	0−2	1−8	55
ITM Power	United Kingdom	0−5	1−80	45−80
AREVA H₂Gen	France	0−10	1−35	45−55
H-TEC System	Germany	0−5	1−30	≈ 55

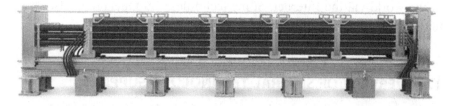

Figure 1.16 Photograph of the 1.25 MW Silyzer 200 PEM water electrolysis stack. Source: Courtesy Siemens Co.

Table 1.9 Main Technical Characteristics of Silyzer 200 (Siemens Co.)	
Electrolysis Type	PEM
Rated stack power	1.25 MW
AC input power	1.9 MVA
Start-up time (from standby)	< 10 s
Operating temperature	60–80°C
Operating pressure	Up to 35 bar
H_2 purity (depends on operation)	99.5%–99.9%
H_2 quality	5.0 (deoxo-dryer option)
H_2 rated production	20.3 kg$\cdot$h^{-1}; 225 Nm^3$\cdot$h^{-1}
O_2 rated production	162.4 kg$\cdot$h^{-1}; 112.5 Nm^3$\cdot$h^{-1}
Design lifetime	> 80,000 h
Dimension skid	6.3 × 3.1 × 3.0 m^3
Skid weight	17 ton
CE conformity	Yes
Tap water requirement	1.5 L$\cdot$Nm^3$_{H2}$$^{-1}$

Fig. 1.17 shows the electrolysis Hall of the Energy Park installed at Mainz, Germany (details about the project can be found online at http://www.energiepark-mainz.de/presse/). There are three Silyzer 200 of 1.25 MW (nominal) or 2.0 MW (peak) power, installed in the electrolysis hall. The idea is to operate the unit at nominal power for baseline hydrogen production, but also to be capable to provide grid services or to respond to extra-production needs, whenever possible or necessary, thus extending the range of operational flexibility of the plant. Usually, operation at peak power is restricted to short periods of time (several tens of minutes on a periodical or occasional basis), because of ageing effects. At such power scale, peak operation is usually restricted by heat removal. In other words, when the outside temperature is low (winter time in nontropical areas), peak operation is easier to implement.

In the electrolysis hall of Fig. 1.17, each of the three electrolysis unit can deliver ≈ 20 kg$_{H2}$$\cdot$h^{-1} ($\cong 160$ kg$_{O2}$$\cdot$h^{-1}) in nominal conditions and ≈ 32 kg$_{H2}$$\cdot$h^{-1} ($\cong 256$ kg$_{O2}$$\cdot$h^{-1}) at peak conditions. This is the largest PEM water electrolysis plant ever built so far (6 MW, 35 bar, 60 kg$_{H2}$$\cdot$h^{-1} $\cong 480$ g$_{O2}$$\cdot$h^{-1}). Hydrogen and oxygen are delivered at pressure up to 35 bar. Mechanical compression is used to deliver hydrogen at 80 bar (first compression stage) or 225 bar (second compression

Figure 1.17 Photograph of the PEM water electrolysis hall of the Energy Park at Mainz, Germany. Source:
Courtesy Siemens Co.

stage for trailers) for compressed storage. The maximum power of the electrolysis plant is ≈ 3.75 MW (nominal) or 6 MW (peak) at system level. The maximum gas production rate of the plant is 60 $kg_{H2} \cdot h^{-1} \cong$ 480 $kg_{O2} \cdot h^{-1}$ in nominal conditions and 96 $kg_{H2} \cdot h^{-1} \cong$ 768 $kg_{O2} \cdot h^{-1}$ in peak conditions. Some performance data can be found in the literature (Kopp et al., 2017).

1.8.2 Proton OnSite

Proton OnSite (Wallingford, United States, http://www.protononsite. com/) is developing scalable $1-2$ MWelectrolyser platforms. Fig. 1.18 shows a 250 kW PEM water electrolysis stack (M-series) that can deliver 50 $Nm^3 \cdot H_2 \cdot h^{-1}$ at the nominal current density of 1.85 $A \cdot cm^{-2}$, nominal operating temperature of 58°C, and nominal pressure of 30 barg. Technical details are listed in Table 1.10.

Fig. 1.19 shows an overview of a containerized PEM water electrolysis units developed by Proton OnSite.

Fig. 1.20 shows a hydrogen fueling station equipped with a PEM water electrolysis unit. The station, supported by the US Department of Energy and the National Renewable Energy Laboratory, is located at Brentwood, DC, and is being hosted by the National Park Service at their Brentwood Avenue maintenance facility for a duration of 4 years. As a collaborative project, Proton OnSite provided the electrolyzer and accessories to enable a fully operational fueling station that

Figure 1.18 Photograph of an M-Series 250 kW PEMWE stack. Source: Courtesy Proton OnSite, a Nel Company.

Table 1.10 Characteristics of the 250 kW Stack of Fig. 1.19A	
Parameter	**Specification**
No. of cells	100
Cell active area	680 cm^2
Hydrogen output	50 Nm^3·h^{-1}
Input power	250 kW
Mass (approximately)	295 kg
Dimensions (approximately)	49 cm W × 61 cm D × 89 cm H
H$_2$ pressure	30 barg
O$_2$ pressure	Ambient
Operating temperature	58°C

will serve FCEV deployment in the metro Washington area, and provide an educational and outreach opportunity to zero emission vehicle stakeholders. The station is available to fuel FCEVs involved in activities associated with the demonstration, while also validating the safety and reliability of PEM electrolysis as a source of zero emission fuel. Though not a public access station, the advancements in sustainable transportation and fueling demonstrated through this project will enable public retail stations of the future.

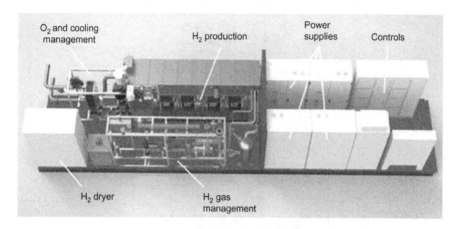

Figure 1.19 Overview of a PEM water electrolysis platform developed by Proton OnSite. Source: Courtesy Proton OnSite, a Nel Company.

Figure 1.20 Hydrogen refueling station equipped with a Proton OnSite PEM water electrolyser fueling the General Motors Chevy Colorado ZH$_2$ fuel cell electric truck developed in collaboration with the U.S. Army . Source: Courtesy Proton OnSite, a Nel Company.

1.8.3 Hydrogenics

Hydrogenics (Canada and Belgium, Oevel, http://www.hydrogenics. com/) is the worldwide leader in designing, manufacturing, building, and installing industrial and commercial hydrogen generation, fuel

Figure 1.21 Photograph of a 1.5 MW electrolysis module (310 Nm³ h⁻¹), 35 bar. Source: Courtesy Hydrogenics Co.

Figure 1.22 Photograph of a short stack with balance of plant . Source: Courtesy Hydrogenics Co.

cells, and MW-scale energy storage solutions. Regarding PEM water electrolysis, 2 MW systems are under testing. Fig. 1.21 shows a PEM water electrolysis stack (maximum input power 1.5 MW, maximum hydrogen output $310\,\mathrm{Nm^3 \cdot h^{-1}}$, maximum operating pressure 35 bar). Fig. 1.22 shows an overview of a short electrolysis module and the BoP. PEM stacks are rated at $2\,\mathrm{A \cdot cm^{-2}}$ nominal current density. Oxygen is released at atmospheric pressure and systems operating under pressure difference are under development.

Figure 1.23 MW-scale PEM water electrolysis plant. Source: Courtesy Hydrogenics Co.

As with many other manufacturers of PEM water electrolyzers, Hydrogenics provides commercial solutions of containerized PEM water electrolysis plants (Fig. 1.23) for various end-users (e.g., chemicals or pharmaceutical industry, electronics and semiconductors, metal processing, power plants).

1.8.4 ITM Power

ITM Power (Sheffield, United Kingdom, http://www.itm-power.com/) manufactures integrated hydrogen energy solutions that are rapid response and high pressure, meeting the requirements for grid balancing and energy storage services, and for the production of clean fuel for transport, renewable heat, and chemicals. Fig. 1.24 shows the photograph of a MW-scale multistack electrolysis module that can be used to accommodate fluctuating power profiles while generating hydrogen at pressures suitable for either direct injection into natural gas networks or via methanation processes without additional compression. Fig. 1.25 shows the photograph of an HGas module, which is a self-contained module. HGas can be used for multiple applications such as clean fuel, energy storage, and renewable chemistry. ITM Power provides a fully integrated turnkey solution, to fulfill customer needs. It is based around a modular platform (standard freight containers) and can be expanded at any point after the initial installation enabling a staged rollout of hydrogen fuel. Table 1.11 provides some specifications of the electrolysis modules. Table 1.12 provides the main characteristics of ITM's Power electrolysis platform that is fully integrated, autonomous, and carries a EU (European Union) mark.

Figure 1.24 Photograph of a multistack PEM water electrolyser. Source: Courtesy ITM Power Co.

Figure 1.25 Photograph of an HGas module equipped with PEM technology. Source: Courtesy ITM Power Co.

1.8.5 AREVA H$_2$Gen

AREVA H$_2$Gen Co. (http://www.arevah2gen.com/fr/societe) is the French leader on PEM water electrolysis. Hydrogen of electrolytic grade is produced from tap water and electricity (grid or renewable energy sources such as solar and wind). Existing markets are mainly in the industry sector where hydrogen is used as chemical feedstock. Emerging markets such as hydrogen refueling stations and energy storage (in combination with grid services) are also addressed. AREVA H$_2$Gen has recently released (during the second China International

Table 1.11 HGas Electrolyzer Specifications	
HGas	Specifications
System power	0.1–100 MW
H_2 production rate	45–40,000 kg•(24 h)$^{-1}$
H_2 pressure	20 bar (50 bar optional)
H_2 purity	99.5%–99.999%
Packaging	ISO containers or indoor installation
Source: *Courtesy ITM Power Co.*	

Table 1.12 Main Characteristics of the HGas Electrolysis Platform		
Demonstrated Benefits		
• Full integration means that all necessary subsystems are included as standard	• Modular structure allows flexibility of sizing using well-proven technology and design	• Dynamic stack clamping technology for variable operation and rapid exchange in the field
• ITM's PEM technology providing leading performance, longevity, and self-pressurization	• Hydrogen purity suitable for use in fuel cell vehicles	• A highly skilled and experienced team from design, delivery, and after-sales support
• Rapid response system enables participation in primary and secondary grid balancing markets	• Sophisticated control system with remote control and condition monitoring features	
Source: *Courtesy ITM Power Co.*		

Hydrogen and Fuel Cell Conference and Exhibition, Beijing, August 28–30, 2017) an innovative concept of 60 MW PEM water electrolysis plant, based on 1 MW (nominal power) stack core technology (Figs. 1.26 and 1.27). The stack can be rated at 2 MW maximum power for limited periods of time. This technology is expected to find application in the petrochemical industry but also in the power-to-gas business, and contribute to a more efficient management of multivector grids. As a sole French manufacturer of PEM water electrolyzers, the ambition of AREVA H_2Gen is to commercialize products for a continuously growing worldwide market. The company is an original equipment manufacturer also involved in project engineering and integration.

1.8.6 H-TEC Systems

H-TEC Systems is a research and production company of the GP JOULE Group based in Lübeck, Germany (http://www.htec-systems.de/en/). Fig. 1.28 shows the photograph of a 30 cell stack from their

Figure 1.26 1–2 MW PEM water electrolysis stack. Source: Courtesy AREVA H₂Gen, Les Ulis, France.

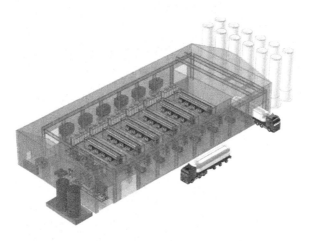

Figure 1.27 Concept of 60 MW PEM water electrolysis plant. Source: Courtesy AREVA H₂Gen, Les Ulis, France.

H-TEC SERIES-S (the reference name is S 30/30). Each individual cell has a 30 cm² active area. The stack technology is also available with 10 and 50 cells. The number of cells in the large stacks is variable. Fig. 1.29 shows the H-TEC Series-ME: ME 100/350 container. The

Figure 1.28 Photograph of a 30 cell S 30/30 stack. Source: Courtesy H-TEC Systems.

Figure 1.29 ME 100/350 PEMWE container. Source: Courtesy H-TEC Systems.

stack contains 140 cells, 450 cm² active area each. It is operated at 30–80°C. The nominal H_2 production rate is 47 $Nm^3 \cdot h^{-1}$ (equivalent to 100 $kg_{H2} \cdot day^{-1}$). Hydrogen is produced with a purity of 5.0 or 99.999% (moist saturated), at a pressure ranging from atmospheric to 30 bar. The nominal energy consumption is 4.9 $kWh \cdot Nm^{-3}$ (74% efficiency).

Fig. 1.30 shows a platform involving a prototype of the H-TEC Series-ME: ME 100/350 container (same specifications as those of

Figure 1.30 Photograph of the ME 100/350 PEMWE module installed in Reußenköge, Germany. Source: Courtesy H-TEC Systems.

Fig. 1.29). The container is located at Reußenköge, north Germany (headquarter of the GP JOULE mother companies, http://www.gp-joule.de). The unit is currently used for on-site testing of PEM electrolysis stacks and to produce hydrogen for a combined heat and power unit, in combination with biogas in their power gap filler (http://www.gp-joule.com/100-renewable/the-power-gap-filler/). In a further step it will be extended to operate as a hydrogen fueling station.

1.8.7 Elchemtech Co.
Elchemtech Co. Ltd is a Korean research and production company based in Seoul, Korea (http://www.elchemtech.com/). Fig. 1.31 shows the photograph of a PEMWE stack. Detailed technical specifications are reported in Table 1.13.

Table 1.14 provides the characteristics of the H_2Gen product line. Hydrogen generators of different size (from 1 to 230 $kg_{H2} \cdot h^{-1}$) are commercially available.

1.8.8 NEL ASA
NEL ASA (Oslo, Norway, http://nelhydrogen.com/) is a global, dedicated hydrogen company, delivering optimal solutions to produce, store, and distribute hydrogen from renewable energy. The company serves industries, energy, and gas companies with leading hydrogen technology. Since its foundation in 1927, NEL has a proud history of development and continual improvement of hydrogen plants. Their

Figure 1.31 Photograph of a five-cell pressurized short stack. Source: Courtesy Elchemtech.

Table 1.13 Elchemtech 3000 cm² Stack Specifications		
Specification	Unit	H₂Gen M-Series
H_2 output(H_2 production rate)	$Nm^3 \cdot h^{-1}$	10
Surface area	cm^2	3000
Cell no.	EA	5
Discharge pressure	bar	350
Power intake(Stack power consumption)	$kwh \cdot Nm^{-3}$	5.5−6
Number of cells	−	5

hydrogen solutions cover the entire value chain from hydrogen production technologies to manufacturing of hydrogen fueling stations, providing all FCEVs with the same fast fueling and long range as conventional vehicles today.

1.9 CONCLUSIONS

PEM water electrolysis has been used for decades for oxygen generation in anaerobic environments and military uses. Regarding civilian applications (hydrogen of electrolytic grade can be used as a raw chemical product or as an energy vector), the technology has now reached the status of "mature technology." The PEM water electrolysis plant, as any other chemical or electrochemical plant, is a complex process and a system that aggregates various subunits. Commercial systems are available at the multi-MW scale and systems are under development and test at the 100 MW-scale. Large-scale deployment

Table 1.14 Elchemtech 3000 cm² Stack Specification						
H₂Gen Specification						
H₂Gen						
Description	PEM Type Hydrogen Generator					
Model	H₂Gen C5	H₂Gen C10	H₂Gen T20	H₂Gen T50	H₂Gen M-Series	H₂Gen G Series
Production (Nm³•h⁻¹ at 0°C, 1 bar)	0.5 Nm³•h⁻¹	1.0 Nm³•h⁻¹	2.0 Nm³•h⁻¹	5.0 Nm³•h⁻¹	10–90 Nm³•h⁻¹	Abobe 100 Nm³•h⁻¹
	1.14 kg• (24 h⁻¹)	2.28 kg• (24 h⁻¹)	4.56 kg• (24 h⁻¹)	11.4 kg• (24 h⁻¹)	22.8–205.2 kg• (24h⁻¹)	228 kg•h⁻¹
Pressure option — Basic	9.9 bar					
Pressure option — Option 1	40 bar					
Pressure option — Option 2	350 bar					
Power consumption	5.5–6 kW•Nm⁻³					
Purity option — Basic	99.95%					
Purity option — Option 1	99.995%					
Purity option — Option 2	99.9995%					
Purity option — Note	Water vapor (5 ppm (dew point: −65°C))					
Coolant	Air cooled:Ambient air, 5–40°C		Liquid cooled:Nonfouling			
DI water requirement	ASTM type II deionized water required, <1 micro Siemen•cm⁻¹ (>1 MegOhm•cm)					
	ASTM type I deionized water recommended, <0.1 micro Siemen•cm⁻¹ (>10 MegOhm•cm)					
Electrical specification	380–480 VAC, three phase, 50–60 Hz					

still requires cost reduction, but investments made to increase the size of commercial system tend to bring cost down and the target €500 kW⁻¹ set by the European Commission in its multiannual work plan will probably be reached in the future.

REFERENCES

Kopp, M., Coleman, D., Stiller, C., Scheffer, K., Aichinger, J., Scheppat, B., 2017. Energiepark Mainz: technical and economic analysis of the worldwide largest power-to-gas plant with PEM electrolysis. Int. J. Hydrogen Energy 42, 13311–13320.

Smolinka, T., Tabu Ojong, T.E., Lickert, E.T., 2015. In: Bessarabov, D., Wand, H., Li, H., Zhao, N. (Eds.), PEM Water Electrolysis for Hydrogen Production: Principles and Applications. CRC Press, Taylor & Francis Group, Boca Raton, FL, pp. 11–33. (Chapter 2).

Key Performance Indicators

2.1 OVERVIEW OF KEY PERFORMANCE INDICATORS USED FOR PERFORMANCE ASSESSMENT

Various key performance indicators (KPIs) can be used to evaluate and compare the performance levels of modern water electrolyzers. They can also be used as performance assessment criteria or used in a roadmap to set target values and implement R&D programs. An example is provided in Table 2.1. This is an adaptation of the "2015 Multi-Annual Work Program of the FCH-JU of the European Commission" (FCH-JU, Fuel Cells and Hydrogen Joint Undertaking, is a public/private partnership between the European Commission the fuel cell and hydrogen industry, and a number of research bodies.) (FCH-JU, 2018). The roadmap up to 2023 shows performance improvements required market applications in the frame of the hydrogen economy.

In Table 2.1, the energy consumption (with units of Joule or kWh per unit of mass or volume) makes reference to the amount of energy (both electricity and heat) of the entire water electrolysis plant (including the energy consumption of the electrolyzer itself, and also the energy consumption of any ancillary equipment). Efficiency (with units of %) refers to the real energy consumption compared with a reference case, which is usually that of the water dissociation reaction under equilibrium conditions. Flexibility is a term that refers to the ability of the water electrolysis plant to operate at any power set point (or current density) between zero (idle) and nominal power, and, in some cases, the ability to run to maximum power. The reactivity of the electrolysis plant tells us how fast the electrolysis plant can move from a given set point to another. Durability makes reference to the lifetime of the various items of equipment of the water electrolysis plant or to the rate of efficiency loss. Safety refers to the maximum hydrogen content found in oxygen due to gas cross-permeation, for various operating conditions of practical interest. Capex or capital expenditure (with units of € per mass or volume) refers to the investment cost over the

PEM Water Electrolysis. DOI: https://doi.org/10.1016/B978-0-08-102830-8.00002-3

Table 2.1 List of Key Performance Indicators Used for Water Electrolysis; 2014 Values, and Improvement Roadmap of the European Commission (KPI = Key Performance Indicator)

			2014	2017	2020	2023
KPI1	Efficiency	Energy consumption (kWh/kg @ nominal power)	57–60 @ 100 kg/day	55 @ 500 kg/day	52 @ 1000 kg/day	50 @ 1000 kg/day
KPI2	Flexibility	Flexibility with degradation < 2% year	5%–100% nominal P	0%–150% nominal P	0%–200% nominal P	0%–300% nominal P
KPI3	Reactivity	Cold and hot start from min to max power	1 min	10 s	2 s	< 1 s
KPI4	Durability	% efficiency loss @ nominal power + 8000 h/year	2%–4%/year	2%/year	1.5%/year	< 1%/year
KPI5	Safety	H_2 content in O_2 @ nominal stationary power	<25% ILE	<25% ILE + flexibility	<25% ILE + reactivity	
KPI6	Capex	Capex with BoP + commissioning	8.0 M€ (t/d)	3.7 M€ (t/d)	2.0 M€ (t/d)	1.5 M€ (t/d)

Adapted from the multiannual work plan of the FCH-JU, 2015–2020

lifetime of the system. Finally, Opex or operational expenditure refers to all operational costs, including mainly the energy cost of the water slitting reaction, but also maintenance costs of all plant equipment. No figure is provided since the energy cost (in € per kWh) can vary quite significantly from one place to another.

Another and more detailed example of the KPIs used to compare the three main water electrolysis technologies is provided in Fig. 2.1. Despite the difficulty associated with building and using quantitative scales that are unanimously recognized by the water electrolysis community, the polar plot provides an easy way to identify, at a glance, where strengths and weaknesses lie. A reference case with 2017 state-of-the-art values was designed and plotted with a dashed line. A first group of KPIs is used to describe the capacity of the equipment. This is mainly the range of operating conditions, which is basically the T, P, and j range (T: temperature, P: pressure, j: current density). The KPIs of the second group are used to measure the level of advancement of the proposed technology, in terms of production capacity, engineering mastering, and manufacturing capabilities. This is mainly the production capacity in terms of active area (that shows the capability to

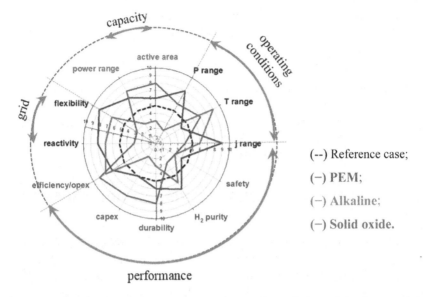

Figure 2.1 Performance comparison of the three main water electrolysis technologies: (--) Reference case; (−) PEM; (−) Alkaline; (−) Solid oxide.

design and put together large cells) and power (that shows the capability to design large systems needed by the market). The KPIs of the third group are used to assess the ability to provide grid services (in the frame of the energy transition, power grids are calling for electrical systems that are flexible, i.e., which can change their power set point between minimum and maximum values, and that are reactive, i.e., which can quickly change power setting). The KPIs of the last group are used to measure various performance levels and to determine the impact of operating conditions on system performance. It should be noted that the list of KPIs and their reference/target values can change with changing market objectives. They can be adjusted on an annual basis, as a function of progress in the field.

The reference case provided in the black dotted lines shows the state-of-the-art of commercial systems in 2017, where:

j range: $0-2.0$ A/cm^2
T range: $25-80°C$
P range: $0-35$ bar
Cell active area: 1000 cm^2
Power range: $0-1$ MW
Flexibility: Stationary at several j values

Reactivity: $0 \rightarrow$ nominal power $<$ minutes
Efficiency @ nominal j: < 60 kWh_{kgH2}^{-1}
Capex: 1500 €/kW @ 1 MW-scale
Durability: $< 5\%$ performance loss/year
H_2 purity at delivery: 4 N
Safety: % H_2 in O_2 $< 25\%$ of the inferior limit of explosivity (ILE)

In the following sections, the KPIs of PEM water electrolyzers only will be considered and discussed.

2.2 OPERATING CONDITIONS AND PRODUCTION CAPACITY

2.2.1 The j Range

The nominal (and in some cases the maximal) operating current density of a proton-exchange membrane (PEM), a water electrolysis cell, or stack of any technology manufacturer indicates how compact the water electrolysis reactor is. It has a direct impact on process Capex and Opex (see below). In the industry, operating current densities of most commercial electrochemical systems (water electrolyzers, and also other large-scale systems, used in the aluminum or chlorine industries, for example) are usually in the $0-1$ A/cm^2 range. The unique features of PEM cells (in particular the small thickness of the cells, the high protonic conductivity of the polymer electrolyte, and the reduced charge transfer overvoltages obtained with platinum group metals (PGMs), tend to shift the upper limit of operating current densities to much higher values. Fig. 2.2 shows typical I$-$V curves measured on various laboratory PEM water electrolysis cells up to 10 A/cm^2 (Millet, 2014; Millet, 2018). Recently, up to $18-20$ A/cm^2 has been reported (Nereng, 2017).

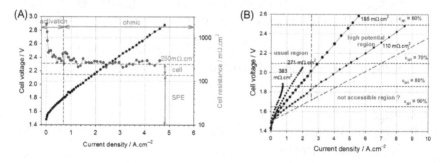

Figure 2.2 (A) Typical IV curve and cell resistance vs operating current density. (B) IV curves of various PEM water electrolysis cells.

Fig. 2.2A shows a typical laboratory-scale IV curve obtained with a Nafion membrane and its first-order derivative with regard to current density. The cell resistance, which is almost infinite at the onset of current, rapidly decreases in the activation domain and then reaches a constant value of $250 \, m\Omega/cm^2$. The latter is the sum of the electrolyte (ionic) resistance (including also the impedance of catalyst layers, CLs) and other cell component (electronic) resistances. Obviously, an effort has to be made to reduce the internal cell resistance and reach higher conversion efficiencies.

Fig. 2.2B shows several IV curves over an extended range of operating current densities. Three distinct domains are observed in such a cell voltage versus current density plot. The first domain is that of usual operating conditions, up to $2.5-3 \, A/cm^2$. The second domain is the high current density domain that can be reached using conventional cell materials and design, but it requires unacceptably high cell voltages, $>2.5 \, V$. In this region, the efficiency is low, heat dissipation high, and durability insufficient for practical applications. The third and last domain is the high current density/low cell voltage domain. This is the most interesting one because it combines a high capacity (of reduced Capex) and high efficiency (of reduced Opex). However, this is the most difficult to reach.

To design a cell capable of operation in this domain of performance, it is necessary (i) to improve the efficiency of the CLs by using more efficient electrocatalysts or by increasing roughness factors, and (ii) to reduce the internal cell resistance by using thinner and better conducting polymer electrolytes. To some extent, this can be achieved using conventional materials, despite the fact that thinner membranes usually increase gas cross-permeation and degrade the safety KPI, especially for operations under pressure. In this regard, significant R&D efforts will be required, with the support of public funding agencies, before breakthrough solutions can be found.

2.2.2 The T Range

The temperature limit of a PEM water electrolysis cell is dictated by the thermal stability of perfluorosulfonic acid (PFSA) materials found in the membrane and the catalytic layers. For conventional Nafion products, this is usually up to $80-90°C$, with a preference for a maximum at $60°C$, to meet durability requirements. Depending upon the

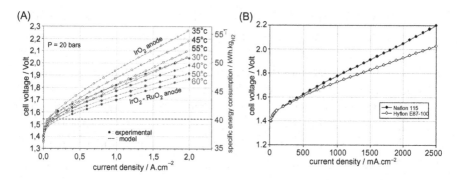

Figure 2.3 (A) Typical IV curves recorded at different operating temperatures using: (●) IrO₂/Nafion117/Pt-C; (●) IrO₂-RuO₂/ Nafion117/Pt-C. (B) IV curves measured with Nafion and Hyflon E87−100 PFSA membranes at 90°C. IrO₂/membrane/Pt-C.

catalyst composition used at the anode for the oxygen evolution reaction (OER), on the technique used for the engineering of CLs, and on the internal cell resistance, performances can differ significantly (Fig. 2.3A).

Fig. 2.3B shows two IV curves measured at 90°C using two different membrane electrode assemblies (MEAs) with the same catalyst compositions but with different polymer electrolyte membranes. The first one was obtained using Nafion 115 (EW = 1100 ± 20 g/eq SO3H; dry thickness = 127 μm) and the second one using Hyflon E87−100 (EW = 870 ± 20 g/eq SO3H; dry thickness = 100 μm). When using Nafion membranes, the maximum operating temperature should be 60−65°C. The closer the temperature is to 100°C, the less durable the membrane is. For temperatures > 100°C (under pressure), membranes lose their mechanical properties and fall apart. However, when using short-side-chain PFSA, the temperature range can be extended up to 130−150°C under appropriate operating pressure (Millet, 2008).

2.2.3 The P Range (Pressurized PEM Water Electrolysis)

Despite its interest as a chemical energy vector, hydrogen is a gas that requires compression for practical storage and transport to customers and end-users. For example, hydrogen mobility using fuel cell vehicles requires compression at or above 700 bar for on-board refueling. Basically, there are three different ways to deliver pressurized hydrogen. The first option is to use an atmospheric water electrolyzer and to compress hydrogen up to the pressure of interest using a mechanical

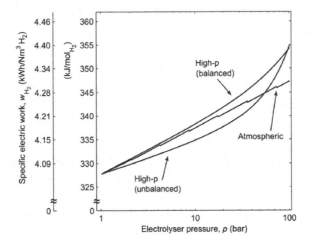

Figure 2.4 Specific energy demand for three different pathways to produce high-pressure hydrogen at T = 60°C: (1) atmospheric electrolysis with subsequent product compression, (2) balanced high-pressure electrolysis, and (3) unbalanced high-pressure electrolysis.

compressor. The second option is to use a balanced high-pressure water electrolyzer that delivers hydrogen and oxygen at the same pressure. The third option is to use an unbalanced high-pressure water electrolyzer that delivers pressurized hydrogen and oxygen at atmospheric pressure. A comparison of the compression energy required by the three different approaches (see Fig. 2.4) reveals that up to a pressure of approximately 40 bar pressure unbalanced water electrolysis requires less energy than the two other approaches (Bensmann et al. 2013; Hanke-Rauschenbach et al., 2016). However, for pressures >40 bar, the energy consumption of the atmospheric electrolyzer coupled to a mechanical compressor becomes less than the energy consumption of pressurized water electrolyzers (e.g., up to 16% at 100 bar compared to in the atmospheric pathway is reported). However, differences are small and arguments other than energetic ones (e.g., overcost associated with the compression vessel, gas purity management) might play a role in deciding which technology is the most appropriate one for a given application.

2.2.4 Active Cell Area

Currently (2017), the surface area of most commercial PEM water electrolysis cells is usually in the range 600–1500 cm^2. Surfaces tend to match the range of DC power sources available on the market that set the maximum current output and hence the maximum current density

(j_{max}). The effort made by technology manufacturers to reach the 10 MW range requires the use of larger cells. In the near future, it can be expected that the standard surface area of PEM water electrolysis cells will become dictated by the commercial availability of PFSA films. The new trend in the industry is to design and use rectangular cells based on the width of commercially available PFSA materials.

2.2.5 Production Capacity

The hydrogen business is calling for water electrolysis plants of increasing capacities. Currently, 1–10 MW range plants are in operation (De Volder, 2015). Despite the fact that the market needs, as related to the energy transition and the large-scale production of hydrogen for energy storage, are still limited, investments are being made to develop commercial systems of larger capacity. According to technology manufacturers, the 10–100 MW range should be reached within a few years. The high flexibility of PEM plants makes the technology attractive in the industrial sector as well, for various applications. This is a significant driver for the development of larger plants.

2.3 EFFICIENCY

2.3.1 Efficiency of the PEM Water Electrolysis Cell and Stack

The efficiency $\varepsilon_d(T, P, j)$ of the water dissociation reaction by electrolysis as well as the total cell efficiency $\varepsilon_{tot}(T, P, j)$ that takes into account the coulombic efficiency $\varepsilon_C(T, p, j)$ of the cell are defined in Chapter 3, Fundamentals of Water Electrolysis, Volume 1, Section 3.4.

2.3.2 Efficiency of the PEM Water Electrolysis Stack

A stack containing N cells of individual surface area A_{cell} can be considered as a single cell of total surface area $N.A^{cell}$ (in cm^2) fed by a DC current (I_{DC} in A) of $N.I_{DC}$ and a DC cell voltage of U^{stack}/N (in V). Hence, the dissociation, current and total efficiency definitions that prevail for a single cell can be easily modified to apply to a N-cell stack. With such a definition, the efficiency of the PEM stack is equal to the mean energy efficiency of the different cells (when the cell voltages of the N cells can be measured individually, it is still possible to calculate the efficiency of each of these cells in the stack).

$$\varepsilon_{tot}^{stack}(T, P, j) = \varepsilon_d^{stack}(T, P, j) \cdot \varepsilon_c^{stack}(T, P, j) \qquad (2.1)$$

Using Eq. (3.23) as the definition of the individual PEM water electrolysis cell, one obtains:

$$\varepsilon_d^{\text{stack}} = \frac{N.U_{\text{rev}}}{U^{\text{stack}}} \tag{2.2}$$

Taking into account the specific energy consumption W_{anc} (J/mol[1]) of the ancillary equipment used to operate the stack (e.g., the energy required to purify and heat the feed water, to pump water through the stack, etc.), one obtains:

$$\varepsilon_d^{\text{stack}} = \frac{N.U_{\text{rev}}}{U^{\text{stack}} + \frac{W_{\text{anc}}}{nF}} \tag{2.3}$$

2.3.3 Efficiency of the PEM Water Electrolysis Plant

As described in Section 1.1, the electrolysis plant contains various energy consuming subunits. Each of them has its own efficiency. Depending on the supplier, similar equipment can have different costs and efficiencies. Therefore, the efficiency of the water electrolysis plant relates the total energy required to produce a given amount (mass or volume) of hydrogen to the reference thermodynamic (equilibrium) case. The best option to calculate this efficiency is to measure the entire energy consumption of each Balance-of-Plant (BoP) component. This is possible when individual energy consumption sensors are placed in individual subunits.

A convenient and practical way to define the efficiency of the PEM water electrolysis plant is to relate the energy content of the hydrogen production to the total energy consumption. Since industrial PEM electrolyzers are not thermostated, the overall energy consumption is the specific electricity consumption measured on the AC/AC transformer. In a derivative form, one obtains:

$$\varepsilon_{\text{plant}}^{\text{HHV}} = \frac{\dot{n}_{H_2}.E_{H_2(T,P)}}{P_{\text{plant}}} \tag{2.4}$$

where:

$\dot{n}_{H_2}$ (mol/s) is the hydrogen molal flow rate at the exhaust of the PEM water electrolysis plant.

$E_{H_2(T,P)}$ (J/mol) is the specific energy content of hydrogen at (T, P) conditions of delivery.

P_{plant} (W) is the power input.

In Eq. (2.4), there is no need to measure the coulombic efficiency or the efficiency of individual BoP units. The efficiency of the plant can be directly obtained by a measure of the energy consumption and the hydrogen molal or mass production. The efficiency of conditioning steps (drying, compression) is also included. The energy content of hydrogen is taken as high (HHV) or low (LHV) heating value, depending on the downstream application.

Due to the electro-intensive character of the water electrolysis reaction, up to 90% of the electricity consumption is taken by the stacks. Fig. 2.5 shows details of the energy flow during stationary operation of a power-to-gas plant used to fill a pressurized truck trailer. Fig. 2.6 shows the energy balance of a PEM water electrolysis plant (Don Quichote project, European Commission, 26-cell stack).

It should be noted here that the efficiency of the PEM plant is constant when the plant is operated under stationary conditions, but efficiency is a function of time when the plant is operated under transient power conditions. This is trivial, however, the calculation of the instantaneous efficiency of the process can show some specific problems, especially because of nonisothermal conditions. For example, Fig. 2.7A shows a plot of the energy consumption at the onset of a rising power set point (this is the case when the electrolyzer is operated

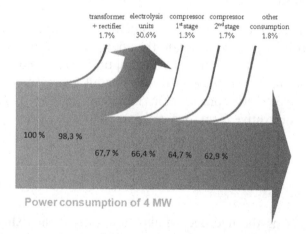

Figure 2.5 Sankey diagram of a 4 MW PEM water electrolysis plant during H_2 filling of a pressurized truck trailer (De Volder, 2015).

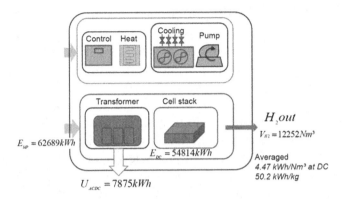

Figure 2.6 Energy balance of a PEM water electrolysis plant (Vaes, 2016).

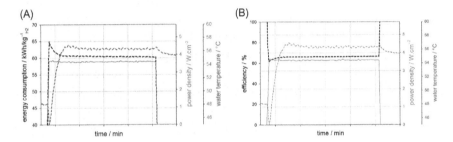

Figure 2.7 (A) Energy consumption during operating in transient conditions. (B) Efficiency during operating in transient conditions.

with transient power cycles) and Fig. 2.7B shows the corresponding transient efficiency.

The lack of thermal regulation generates stack temperature changes. There is an overconsumption of energy (and a corresponding process under-efficiency) just before power plateaus until internal dissipation brings the operating temperature back to its nominal value.

2.4 FLEXIBILITY–REACTIVITY

2.4.1 Definitions and Objectives

Most modern water electrolyzers are designed for operation under stationary conditions, i.e., constant (j, T, P) conditions, at nominal current densities (j), on an 8000-h/year basis. Because up to 80% of the hydrogen cost is due to the cost of kWh, cost-competition requires the use of low cost kWh. In the frame of the energy transition, as a result of the increasing share of intermittent energy sources in the energy mix, the conditions of power grid management are changing. New service opportunities to access less costly kWh are appearing. Hence, future markets of electrolytic hydrogen (e.g., large scale plants for energy management or hydrogen refueling stations) will require more flexible electrolysis plants, which can operate at subnominal and supnominal operating conditions. Flexibility is a general term used to measure how flexible a water electrolyzer is. More specifically, flexibility is a term that describes the ability of the PEM water electrolysis plant to operate at different input power levels, between zero and maximum power. Future markets of electrolytic hydrogen will also require more reactive technologies. Reactivity is a term used to measure the ability of the PEM water electrolysis plant to change from one power set point to another one in an usually limited amount of time. Reactivity can be measured at cell or stack level, but also, and most importantly, at system level. System reactivity depends on stack capability and on BoP design (e.g., pressure management) and downstream requirements (e.g., compression or purification). More specifically, reactivity is a measure to the time required by the electrolysis unit to respond to power constraints and reach subnominal/supnominal current densities (or any specific operating powers). Power grid management calls for electrical technologies capable of switching from idle state (zero net power consumption) up to nominal power, and vice versa, within a few seconds only. Fig. 2.8 shows examples of power profiles of interest for both primary and secondary reserve management (Allidières et al., 2018). This is basically a succession of power ramp-up and ramp-down profiles. The slopes in the transient sections and the duration of the power plateaux are dictated by national power grid regulations and can differ from one country to another.

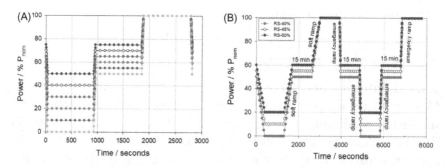

Figure 2.8 (A) Load profiles of interest for the management of the primary reserve. (B) Load profiles of interest for the management of the secondary reserve.

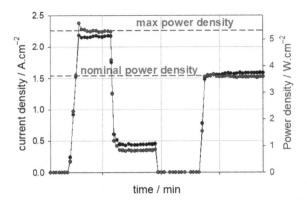

Figure 2.9 Current and power flexibility of a PEM water electrolyzer at system level.

2.4.2 Performances

Flexibility and reactivity can be both measured either on the individual PEM cell, or on the PEM stack, or on the PEM plant. The next figure (Fig. 2.9) shows examples of how flexible PEM water electrolyzers can be at system level. Usually, PEM electrolyzers are operated in intentiostatic mode. The figure shows full flexibility in terms of current density and power density. The hydrogen production unit can stand idle and then jump to either nominal or maximal current density (power) within seconds, depending on which power reserve the plant is qualified for.

Of course, the efficiency of the water dissociation reaction will differ if this is a cold start (the electrolyzer is at ambient temperature) or a hot start (the electrolyzer is close to nominal operating temperature). At the MW-scale, this can make a significant difference. In general, when PEM electrolyzers are operated intensiostatically and nonisothermally, this leads to stack voltage overshoots and, as a result, to power overshoots. Fig. 2.10 shows an example of the situation. The stack voltage overshoot decreases as the system temperature reaches stationary conditions. Therefore, strategies are required to either maintain the nominal temperature, or to buffer heat and transfer it back to the electrolyzer before power can be turned on.

Power fluctuation during stationary operation is another KPI of interest for grid services. Specifications may vary from one grid to another but qualification for grid services usually requires percent-range power stability. Fig. 2.11A shows an example measured on a

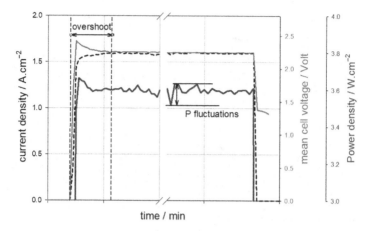

Figure 2.10 Current density, mean cell voltage, and power density measured on a large PEM water electrolyzer during a power on/off experiment.

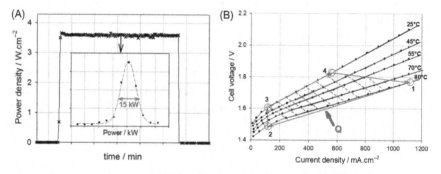

Figure 2.11 (A) Power fluctuations and dispersion during stationary operation. (B) Hysteresis cycle during non-isothermal power on/off operation.

power PEM stack: There is a 4% magnitude of power fluctuation on the power plateau, corresponding to a Gaussian mid-height dispersion of 15 kW. Regarding nonisothermicity during power cycling, the situation can be appreciated from Fig. 2.11A where the hysteresis cycle followed by the PEM cells when the power is switched on and off is shown. The starting point (1 in the figure) is at nominal operating conditions (80°C, 1.1 A/cm^2). When power is turned off, the cell voltage goes down along the I–V curve at 80°C, down to point 2. At rest time, the cell cools down by heat exchange with the surroundings and the operation point shifts gradually from 2 to 3 (ambient temperature). When power is turned on again, the cell voltage increases up to the

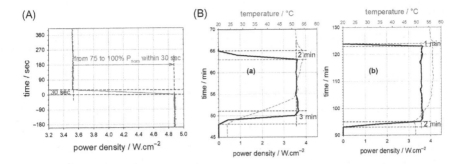

Figure 2.12 (A) Reactivity of a PEM water electrolysis stack for primary reserve. (B) Reactivity of a PEM water electrolysis stack for secondary reserve: (a) cold start; (b) hot start.

nominal current density along the I−V curve at 25°C. Gradually, heat dissipations raises the internal temperature until the initial (j, T) set point is met again. The surface area along 1, 2, 3, and 4 is a heat power density term in W/cm^{-2}: This is the heat dissipation required to go back to the initial set point. It should be noted here that national grid power regulations also provide constraints in terms of power fluctuations during stationary operation.

Regarding the reactivity of PEM water electrolyzers, the next figure shows the situation for both secondary and primary reserves. For primary reserve (Fig. 2.12A), the constraint imposed by grid regulation is to switch power density (in this example from 75% to 100 % of nominal power density) within less than 30 s. For secondary reserve (Fig. 2.12B), the constraint is to switch power density from initial to final power density set point in less than 3 min. During cold start operation, the power set point is reached within 3 min ($\Delta T = 25°C$) and the system is switched off within 2 min. During hot start operation, the power set point is reached within 2 min ($\Delta T = 7°C$) and the system is switched off within 1 min. MW-scale PEM water electrolysis plant are qualified for both reserves.

MW-scale PEM water electrolysis (at stack and BoP levels) have already demonstrated the appropriate level of flexibility and reactivity to satisfy most stringent requirements for power grid operation. This is the reason why significant investments are currently made to increase the power capacity of commercial products in order to install the technology as a key player in the frame of the energy transition.

2.5 SAFETY

2.5.1 Fundamentals of Gas Cross-Over

Safety is a measure of the ability of the water electrolyzer to produce hydrogen according to a set of safety criteria. The two main types of hazards in a PEM water electrolysis unit are electrical hazards and explosive hazards. During operation, cross-permeation of gaseous reaction products (H_2 and O_2) tends to reduce the coulombic efficiency (see section on efficiency) of the electrolysis cell, however, it also raises safety issues. Details about the underlying microscopic phenomena have been analyzed in Chapter 5, Gas Permeation in PEM Water Electrolyzers, Volume 1. Simple models have been reported in the literature to account for cross-permeation (Fateev et al., 2009; Grigoriev et al., 2011; Hanke-Rauschenbach et al., 2016). The mole percentage (mole %) of hydrogen in oxygen is a function of various physical properties of the polymer electrolyte and of various operating electrolysis parameters:

$$n_{H_2}(\%) = \frac{2FD_{H_2}\Delta P_{H_2}}{\left(2FD_{H_2}\Delta P_{H_2}\right) + \left(jLH_{H_2}\right)} \times 100 \qquad (2.5)$$

where:

n_{H_2} (%) is the mole fraction of hydrogen in oxygen in stationary operating conditions.

F (96,485.33 C/mol[1]) is the Faraday constant.

D_{H_2} (in cm^2/s) is the hydrogen diffusion coefficient in the hydrated polymer at (T, P) conditions of operation.

ΔP_{H_2} (Pa) is the pressure difference set across the membrane during operation.

j (A/cm^2) is the operating current density.

L (cm) is the membrane half-thickness.

H_{H_2} (mol/cm^3/Pa[1]) is the solubility coefficient (Henry's constant) of H_2 in the polymer electrolyte.

2.5.2 In Stationary Operating Conditions

Fig. 2.13 shows a plot of the hydrogen percentage in oxygen as a function of operating current density, obtained at different temperatures during PEM water electrolysis operating under a pressure difference of 20 bar. There is good agreement between the experimental (●) data

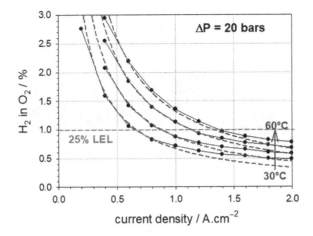

Figure 2.13 Experimental (●) and calculated (--) relationships between % H₂ in O₂ and operating current density.

and calculated (---) values, obtained using Eq. (2.5). This shows that Eq. (2.5) contains all the necessary physical parameters to account for the hydrogen cross-permeation process. In terms of safety, 25 % of the Lower Explosion Limit (LEL) is usually used as a threshold value to evaluate risks during electrolysis. Fig. 2.13 shows that the hydrogen content in oxygen is significant over almost the entire range of operating current densities. At such pressure (20 bar), the situation becomes critical for $j < 0.2$ A/cm^2 because the LEL is reached. Of course, the gas production rate is lower at low current densities so what matters is the amount of energy stored in the plant in the form of H_2/O_2 mixtures. This is why the amount of gas stored in the plant at any time (within the stack but also in the liquid—gas separation units and other BoP units) is the critical factor that determines the level of risk (Millet et al., 2012).

2.5.3 In Transient Operating Conditions
The hydrogen percentage measured in oxygen follows power input (Fig. 2.14). In transient operating conditions, the amount of hydrogen fluctuates and should be closely monitored on an adequate timescale to manage associated explosive hazards.

2.5.4 Mitigation Measures
Despite the remarkable physical properties of PFSA materials, hydrogen cross-permeation remains a critical problem for PEM water

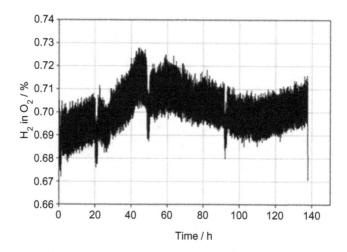

Figure 2.14 H_2 in O_2 fluctuations measured during input power fluctuation.

electrolyzers. Appropriate risk mitigation solutions are required to avoid the explosion of hydrogen/oxygen mixtures. There are two main options. The first one is to incorporate gas recombiners within the cell itself. This is not necessarily a complex and expensive process. For example, isolated platinum particles placed within the membrane are actively promoting the catalytic recombination of hydrogen and oxygen into water. Such particles can be incorporated by impregnation of the membrane with appropriate platinum cationic precursors followed by in situ chemical reduction using a chemical reducer such as sodium borohydride ($NaBH_4$) or any equivalent. Another option is to reduce the hydrogen permeability of the membrane: see Chapter 5, Gas Permeation in PEM Water Electrolyzers, Volume 1, for additional details. Looking at Eq. (2.5), this can be achieved by either changing the operating (T, P) conditions (cross-permeation is less significant at lower operating temperature and pressure but pressurized operation is a key advantage of PEM water electrolysis that should not be given up), or the geometry of the membrane (but thicker membranes also means larger resistances and lower efficiencies so there is no real interest to use thicker membranes, except for some specific applications, when a loss of efficiency is acceptable) or the hydrogen solubility (dictated by membrane properties that also impact conductivity) or the hydrogen diffusivity (interactions with the chemical structure of the membrane; the use of mechanical membrane reinforcements such as polytetrafluorethylene (PTFE) or polybenzimidazole (PBI) nets have positive effects).

2.6 DURABILITY

As for any technology, modern water electrolyzers must be reliable (reproducibility of performances) and durable (ability to maintain optimal performances overextended periods of time). Hence, reliability is a measure of how reliable the technology is and durability is a measure of how long the performances can be maintained. Both aspects are usually covered by the KPI on durability. This is a critical KPI which will, ultimately, tell what is the cost of the technology and what return on investment can be expected. Usually, the durability of electrolysis cells is the most critical factor. The cell voltage at constant current density tends to increase with time (an example is provided in Fig. 2.15). Performance losses differ from one cel design to another, and from one technology supplier to another. Fig. 2.16 shows the results of a long-term test performance on a laboratory cell with periodic cleaning. The long-term drift of only 16 μV/h over 20,000 h of operation demonstrates that the MEAs of PEM water electrolyzers can be operated over long periods and they are not necessarily the weakest part of the plant. Using an appropriate internal reference electrode, it is possible to separately measure anode and cathode voltage drifts (Fig. 2.16). The anode degradation rate (in this example, the individual contributions of catalyst site decrease and membrane corrosion are not differentiated) is usually the fastest one.

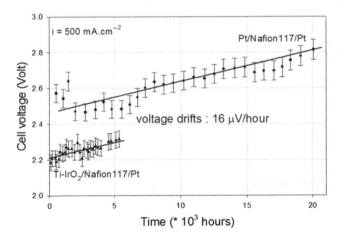

Figure 2.15 Cell voltage vs time of (A) Pt/Nafion 117/Pt and (B) IrO₂/Nafion 117/ Pt cells at a constant current density of 0.5 A/cm² (Millet et al., 1989).

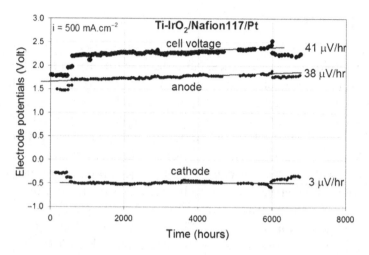

Figure 2.16 Cell voltage, H_2 and O_2 overvoltages measured on a PEM water electrolysis cell operating at a constant current density of 0.5 A/cm² (Millet et al., 1996).

2.7 CAPEX/OPEX ANALYSIS

The technology cost (in €/kW) is a measure of the investment cost for a given production rate. The Capex (in $€/kg_{H_2}$, €/kg) is a measure of the investment cost required for the production of a unit mass of hydrogen (or oxygen) of electrolytic grade. It is usually difficult to obtain costs from individual manufacturers because, as for any product, commercial costs depends on many factors. Therefore, cost analysis usually relies on estimated Capex values. The Capex is a function of the operating current density j. Integration over the lifetime of the electrolysis unit (assuming that the lifetime is not affected by the value of the current density of operation, as long as this value remains within the nominal specification of the technology manufacturer) yields the hydrogen and oxygen cost in $€/kg_{H_2}$:

$$\text{Capex}(€/kg_{H_2}) = \frac{F}{j.10^{-3}.T(s).S(cm^2)}.\text{cost}(€/kW) \qquad (2.6)$$

Opex (in $€/kg_{H_2}$) is a measure of the production costs (mainly energy and maintenance costs) required for the production of hydrogen and oxygen of electrolytic grade. The Opex is a function of the operating current density j:

$$\text{Opex}(€/kg_{H_2}) = \frac{U_{cell}F}{3600}.\text{cost}(€/kWh) \qquad (2.7)$$

Opex is affected by operating conditions (transient modes of operation are expected to increase costs), system durability, and strategies implemented for operation in transient conditions. The total hydrogen/oxygen cost in €/kg is the sum of the Capex (investment, excluding civil engineering cost) and the Opex(energy and maintenance costs). A typical example is provided in Fig. 2.17, assuming a technology cost of €1,500/kW, an electricity cost of €150/MWh, and a lifetime t of 20,000 h.

Regarding Capex, there is a clear exponential decrease of the total cost with operating current density. This is obviously because the size (and hence the cost) of the production unit depends strongly on current density. At low j values, the electrolyzer will produce a limited amount of gases during its lifetime, while at high current density, significantly more will be produced. The Opex dependence on current density arises from the current–voltage characteristics of the PEM cell. The cost increases logarithmically at low current density and then linearly at higher current densities because of internal power dissipation in the cells. Therefore, Capex reduction requires an increase in operating current density while, simultaneously, Opex reduction requires a decrease in operating current density. As a result of these two opposite trends, there is an optimal current density value which, of course, depends mainly on the cost of kWh of electricity. The higher the cost of electricity, the narrower the range of current density of interest, and vice versa.

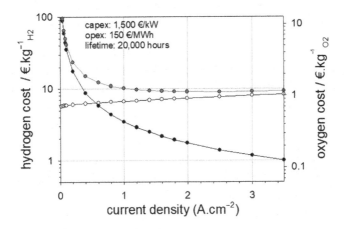

Figure 2.17 Capex, Opex and total hydrogen/oxygen costs as a function of operating current density for a reference PEM water electrolysis plant.

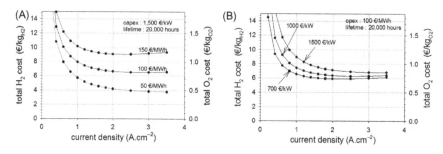

Figure 2.18 (A) Total H₂/O₂ specific cost vs current density for different electricity costs (PEM). (B) Total H₂/ O₂ specific cost vs current density for different Capex (PEM).

The total H_2/O_2 cost vs current density is plotted for different electricity costs in Fig. 2.18A and for different Capex in Fig. 2.18B. The particular sensitivity of the total gas cost to the price of electricity appears clearly. This is not a surprise for electrochemical technologies.

The following figures show cost comparison of main water electrolysis technologies. This is the current situation prevailing in 2017. In the alkaline cell, the Capex is low ($\sim 600-700$ €/kW). Hence, the operating current density is usually low. In the PEM cell, the Capex is higher (~ 1500 €/kW), so it is more interesting to perform electrolysis at several A/cm². In the solid oxide cell, the Capex is even higher ($\sim$ €4,000/kW), and much higher current densities are required to make the technology cost competitive. Although the efficiency is higher, this is strongly counterbalanced by a larger Capex, because this is a more recent technology and because of the ancillary equipment required for steam generation and heat management (Fig. 2.19).

The energy consumption depends on the I−V curve and the operation point that is selected. The alkaline process is favored for its low operating current density (which means high efficiency) and low Capex (because cheap cell materials are used). The PEM process is favored for its possibility to increase the operating current density to reduce Capex while maintaining a high efficiency, compared to the alkaline process. The solid oxide process is favored for its higher efficiency but is negatively impacted by still high Capex. Ultimately, what makes the difference between the three main water electrolysis technologies is: (i) the Capex expressed in €/kW (including durability and maintenance), (ii) the electricity consumption (efficiency), and (iii) the cost of electricity.

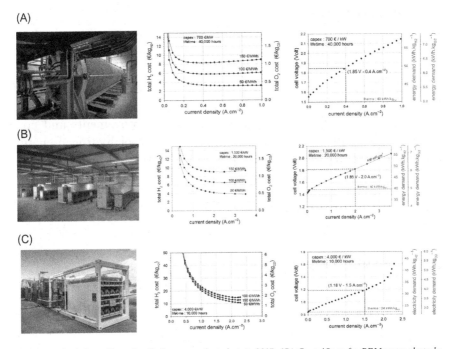

Figure 2.19 (A) Capex/Opex for alkaline water electrolysis in 2017. (B) Capex/Opex for PEM water electrolysis in 2017. (C) Capex/Opex for steam water electrolysis in 2017.

2.8 A SWOT ANALYSIS OF PEM WATER ELECTROLYSIS

In this section, a SWOT (Strengths, Weaknesses, Opportunities, and Threats) analysis of the PEM water electrolysis technology is provided from an industrial perspective.

2.8.1 Main Process Characteristics

PEM water electrolysis is the second most mature water electrolysis technology available on the market. PEM water electrolysis cells contain a thin ($\sim 150-220$ μm thick) proton-conducting polymer membrane. The main advantages are the following: (i) no liquid electrolyte, no electrolyte leakage, and reduced corrosion issues; (ii) high compactness; possibility to operate the cells at elevated current densities of several A/cm^2; (iii) very flexible (cold and hot starts) and reactive (capable of quick power set point changes) for operation with renewable energy sources; (iv) possibility to operate under pressure (up to 200 bar) and under a pressure difference, if required.

2.8.2 Process Maturity

PEM water electrolysis is less mature than alkaline water electrolysis in terms of track records; however, in terms of the production capacity of commercial systems, the difference tends to disappear ($\sim 20-40$ kg_{H_2}/h). Furthermore, PEM technology is more expensive in terms of Capex and the feed water must have a higher purity. However, PEM has a greater potential in terms of technical improvements and cost reduction. The interest of PEM water electrolysis for hydrogen energy applications (grid services and mobility) is currently under investigation at the multi-MW scale level.

2.8.3 Technical Characteristics

Some technical characteristics are listed in Table 2.2.

2.8.4 Economic Characteristics

Some economic characteristics are listed in Table 2.3.

2.8.5 SWOT Analysis

1. **Strengths**
   - Mature technology at the $1-5$ MW scale, with increasing track record.
   - High purity gas delivery (99.999%).
   - High compactness: Commercial operation at $3.0-3.5$ A/cm^2; demonstration at the laboratory scale at $10-15$ A/cm^2.
   - High energy efficiency (70%–80% HHV at 1 A/cm^2) depending on operating current density and temperature.
   - Durability $> 60,000$ h of continuous operation demonstrated in the Navy.
   - Increasing capacities commercially available (systems up to $7-10$ MW available on the market; development up to 100 MW underway).
   - Excellent flexibility and reactivity at system level for operation with transient power sources.
   - Operating under pressure (200 bar demonstrated, 350 bar prototypes) or under a pressure difference (usually oxygen released at atmospheric pressure while compressed hydrogen is delivered).
2. **Weaknesses**
   - For different reasons, the Capex of PEM technology (1700–2500 €/kW for 300–500 kWe systems) is still higher than the Capex of alkaline technology (900–1700 €/kW for 300 kWe–5 MWe

Table 2.2 Set of Technical Characteristics of PEM Water Electrolyzers

Input Specifications

Water specifications	Highly deionized water required (> 10 MΩ/cm). The water purification unit is usually incorporated to commercial plants.
DC specifications	Unknown long-term effects of minor DC signal fluctuations on performance and durability.

Process Products

Co-products	• Oxygen contains H_2O, H_2, and N_2 traces. The level of purification is dictated by application and downstream requirements. • Hydrogen contains H_2O, O_2, and N_2 traces. The level of purification is dictated by application and downstream requirements.
H_2 and O_2 purity	Process products are released at $> 99.9\%$ purity. Main process impurities are H_2O, O_2). Two purification steps are usually provided: • Primary purification: Gases are cooled down to $3-15°C$. A chiller module is used to lower dew-point temperature and remove water. • Secondary purification: Additional purification technology can be added depending on downstream process requirements.
Product conditioning before transport	Electrochemical compression is available (output pressures ranging from 15 to 80 bar, depending on applications). Delivery of higher pressures requires mechanical gas compression. PEM can operate under pressure difference (when O_2 needs to be released at atmospheric pressure).

Operating Conditions and Process Characteristics

Operating temperature	Commercial: $50-70°C$; R&D: up to $150-200°C$ using alternative polymer electrolytes such as short side-chain PFSA or PBI.
Operating pressure	Commercial: $1-80$ bar; R&D: $1-350$ bar.
Production capacity	1 kW-2 MW corresponding to 66.67 Nm^3 H_2/h per unit.
	Demonstration up to 7 MW.
	R&D on $10-100$ MW at system level.
Energy consumption	Thermodynamics: 40 kWh$_e$/kg $H_2 = 3.57$ kWh/Nm^3 H_2.
	Stack: $4.3-4.5$ kWhe/Nm^3 H_2 product @ $0.8-1$ A/cm^2 (enthalpic efficiency of $\sim 79-83\%$ HHV).
	Up to $4.9-5.4$ kWhe/Nm^3 H_2 (equivalent to $66\%-73\%$ efficiency) at system level.

FCH-JU roadmap		2014	2017	2020	2025
	kg/day	100	500	1000	1000
	(kWh/kg)	57–60	55	52	50
	(kWh/Nm^3)	5.2–5.5	5.0	4.8	4.6
	€/kW	2500	1500	1000	800

Process flexibility	High flexibility: operating power can vary between 0% and 100% of max power.
Process reactivity	Reaction time to power changes: (i) hot start: within a few seconds; (ii) cold start: within a few seconds (no need for preheating).
Heat cogeneration	@ process level: thermal regulation during operation produces a water flow at $75°C$ that can potentially be used for heating purposes.
	@ product level: hydrogen can be use in a cogeneration unit (fuel cell, boiler, etc.). LHV = 119.93 MJ/kg.
Lifetime	$60,000$ h at 3.5 A/cm^2 demonstrated in the US Navy.
	$15,000-20,000$ h in stationary conditions with $<10\%$ performance losses (guaranteed by technology suppliers).

Table 2.3 Set of Economic Characteristics of PEM Water Electrolyzers

Capex

Capex range	1700−2500 €/kW for 300−500 kWe systems. Capex breakdown (the relative share of the electrolysis stack increases with production capacity): Electrolysis cells ∼ 45% Power electronics ∼ 15% Ancillary equipment ∼ 25% Power grid interface ∼ 5% Manufacturing, transport & onsite installation ∼ 7% Process engineering ∼ 3%
Capex reduction perspectives	• Increase in operating current density (1.0 → 3.5 A/cm^2) to reduce the cost of the stack; some systems developed for military uses are reported to operate at 3.5 A/cm^2. • Increase in active surface area to reduce cell number. • Reduction of power consumption in AC/DC converters. • Development of larger capacity units using the same approach as those used in the main electrolysis processes. • Less significantly, development of innovative materials and material processes for cell components. • Replacement of PGM electrocatalysts.
FCH-JU roadmap	Toward the cost of alkaline technology.

	SOA	2017	2020	2025
kg/day	100	500	1000	1000
€/(kg/day)	8000	3700	2000	1500
€/kW	3400	1600	920	720

Opex

Opex range	• 55−65 kWh/kg$_{H_2}$ kg (equivalent to 7−8 kWh/kg$_{O_2}$ kg). • The relative contribution of OPEX to total gas cost increases when production capacity increases (up to 80% for > MW-scale).
Opex reduction perspectives	• Fixed Opex (without cell replacement) ∼ 1%−1.5% of Capex • Cell replacement cost ∼ 1060 €/kWe.

Markets

Main markets	• Same as alkaline (chemical industry + mobility + energy storage).
Market volumes	• Grid services.

Product Price

On-site water electrolysis	• Can be installed and used onsite for local gas consumption.

systems). This is mainly due to expensive cell components and non-automated manufacturing and assembly procedures. In the long term, expensive PGMs will also have to be replaced by cheaper materials to meet target Capex costs.

• Opex in kWh/kg$_{H_2}$ (common to all water electrolysis technologies): The energy cost is mainly due to the cost of electricity that dictates the cost of gas produced by electrolysis.

- Safety issues (common to all water electrolysis technologies): Management of hydrogen—oxygen stoichiometric gas mixtures, gas cross-contamination during operation.

3. **Opportunities**
   - The increasing competition with incumbent water electrolysis technologies (alkaline and solid oxide) is boosting investments in R&D.
   - Operation at current density between 1 and 2.5 A/cm^2 is the standard in the industry; systems operating at even higher current densities (3.5 A/cm^2) have already been developed by technology manufacturers; much higher current densities (10 A/cm^2) have also been reached.
   - The availability of polymeric proton conductors for operation up to 250°C would significantly contribute to the reduction of the energy consumption. Significant progress has been made recently with short side-chain PFSA and PBI materials.

4. **Threats**
   - Lack of competitiveness compared to steam methane reforming. There are different national directives that provide safety and legal framework and guidelines for the production of hydrogen. In France, hydrogen production must satisfy ICPE 1415 requirements, irrespective of the technology involved. This regulation framework is evolving, to facilitate the deployment of systems of limited capacity ($<$100—200 Nm^3 H_2/h). Industrial equipment must comply with inhomogeneous international legal frameworks.
   - There is currently an increasing competition between the three main water electrolysis technologies. Alkaline is considered as the most mature and cheapest one, PEM is considered more appropriate for large size systems operating in flexible conditions, while solid oxide is more suited for operation in reversible modes for water-CO_2 coelectrolysis. There is no convincing technical indication to anticipate which technology will dominate in the future.

2.9 CONCLUSIONS

PEM water electrolysis is an efficient technology to produce hydrogen and oxygen of electrolytic grade from water. In state-of-art technology, $\sim$80% HHV efficiency at 1 A/cm^2, equivalent to 55—65 kWh/kg_{H_2}kg (or 6.9—8.1 kWh/kg_{O_2}kg), is commonly achieved. The lifetime of a commercial electrolysis plant is usually in the 15,000—20,000 h range of continuous operation. Some suppliers report 80,000 h. The delivery pressure at the exhaust of the electrolyzer is usually within the 15–20 bar range. To reach the target pressure for automotive applications (e.g., 700 bar for on-board refueling), mechanical compression is required.

REFERENCES

Allidières, L., Brisse, A., Millet, P., Valentin S., Zeller, M. 2018. Test protocols for the qualification of MW-scale PEM water electrolyzers in view of grid services, Proc. 9th ICH2P Conference, Zagreb, Croatia, July 16–18.

Bensmann, B., Hanke-Rauschenbach, R., Pena Arias, I.K., Sundmacher, K., 2013. Energetic evaluation of high pressure PEM electrolyzer systems for intermediate storage of renewable energies. Electrochim. Acta 110, 570–580.

De Volder, M., Energy Park Mainz, a project for the industry. FCH-JU stakeholders forum, 19/11/2015. http://www.fch.europa.eu/sites/default/files/Mainz%20a%20large%20scale%20wind-H%20project%2C%20M.%20de%20Volder%2C%20SF%202015%20%28ID%202848756%29.pdf.

Fateev, V., Grigoriev, S., Millet, P., Korobtsev, S., Porembskiy, V., 2009 High pressure PEM water electrolysis and corresponding safety issues, Proceeds 3rd International Conference on Hydrogen Safety; 16–18 September 2009; Ajaccio, France.

FCH-JU (Fuel Cell and Hydrogen Joint Understanding). http://www.fch.europa.eu/ (Accessed on March 30, 2018)

Grigoriev, S.A., Porembskiy, V.I., Korobtsev, S.V., Fateev, V.N., Auprêtre, F., Millet, P., 2011. High pressure PEM water electrolysis and corresponding safety issues. Int. J. Hydrogen Energy 36, 2721–2728.

Hanke-Rauschenbach, R., Bensmann, B., Millet, P., 2016. Hydrogen production using high pressure water electrolysers. Chapter 10. In: Basile, A. (Ed.), Handbook of Hydrogen Energy. WoodHead Science Editions, London.

Millet, P., 2008. Final report of the GenHyPEM FP6 research project. FCH-JU, European Commission, Brussel, Belgium.

Millet, P., 2014. HydroPEM project final report. Agence Nationale de la Recherche, France.

Millet, P., Durand, R., Pinéri, M., 1989. New solid polymer electrolyte composites for water electrolysis. J. Appl. Electrochem. 19, 162–166.

Millet, P., Andolfatto, F., Durand, R., 1996. Design and performances of a solid polymer electrolyte water electrolyser. Int. J. Hydrogen Energy 21, 87–96.

Millet, P., de Guglielmo, F., Grigoriev, S.A., Porembskiy, V.I., 2012. Cell failure mechanisms in PEM water electrolyzers. Int. J. Hydrogen Energy 37, 17478–17487.

Millet, P., 2018. On the efficiency of PEM water electrolysis cells operating at elevated current densities, Proc. 9th ICH2P Conference, Zagreb, Croatia, July 16–18.

Nereng, L., 2017. Abstract. 1st International Conference on Water Electrolysis, 2017 June 12–15; Copenhagen, Denmark.

Vaes, J., 2016. Field experience with Hydrogenics' prototype stack and system for MW PEM electrolysis. 2nd Int. Workshop on Durability and Degradation Issues in PEM Electrolysis Cells and their Components. 2016 February 17; Freiburg, Germany.

CHAPTER 3

Performance Degradation

3.1 INTRODUCTION

3.1.1 The Ideal Proton-Exchange Membrane Cell

Fig. 3.1 shows a typical design of a proton-exchange membrane (PEM) water electrolysis cell (this is Fig. 4.1 from Chapter 4, The Individual Proton-Exchange Membrane Cell and Proton-Exchange Membrane Stack, Volume 1 repeated here to support the discussion). The distribution of electric potential across the cell, from the anode on the right side of the figure down to the cathode on the left side is shown qualitatively in red. Electrons are the charge carriers in the metallic cell components and hydrated protons are the charge carriers across the polymer electrolyte. Most voltage drops are in ohmic terms (resulting from bulk and contact resistances). The only nonohmic contributions are the two charge transfer overvoltages across the two catalytic layers (2 and 2′).

Let us introduce the concept of an ideal PEM water electrolysis cell. This is not a theoretical cell, this is a "real" PEM cell containing optimized cell components. Its performances are those measured at Beginning-of-Life (BoL), and they are maintained endlessly. This is basically a Pt-C/Nafion-115/IrO_2 cell with Ti porous transport layers (PTLs) or Ti and carbon PTLs. In particular, this ideal PEM cell has the following characteristics:

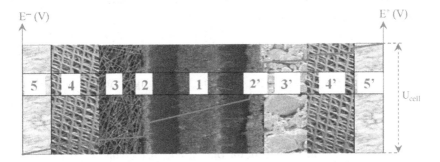

Figure 3.1 Schematic representation (cross-sectional view) of a PEM water electrolysis cell. 1 — PTFE-reinforced PFSA membrane; 2/2′ — catalytic layers (CLs); 3/3′ — porous transport layers (PTLs); 4/4′ — spacers and flow disrupters; 5/5′ — end-plates (Rozain and Millet, 2014).

PEM Water Electrolysis. DOI: https://doi.org/10.1016/B978-0-08-102830-8.00003-5

- A high efficiency (80% @ 1 A/cm^2) and the IV curve of best 2017 practices.
- A high gas purity (4 N @ 50–60 bar).
- The ability to maintain this high level of efficiency indefinitely, without degradation.
- It can operate efficiently under transient power loads, with no impact on gas purity and performance.
- It is scalable in size without any significant impact on the efficiency and durability.
- It can be stacked up to several hundred cells without any significant impact on efficiency and durability.

3.1.2 The Real Proton-Exchange Membrane Cell

The real PEM water electrolysis cell is not an ideal cell. Degradation phenomena do occur, inevitably leading to performance losses during operation. The performance level at BoL is different from the performance level at the End-of-Life (EoL). Cell ageing is observed experimentally. Because the purpose of the PEM water electrolysis cell is to electrolyze water, the I–V curve is a good indicator of ageing phenomena. I–V curves provide useful in situ information on where degradation occurs within the cell, in terms of the degradation mechanism and rate of degradation. Alternatively, chrono-amperometric (CA) experiments can also be used to analyze cell degradation. The cell voltage drift (usually in the 1–20 µV/h range) is commonly used as a measure of the ageing rate (e.g., a cell voltage drift of 5–10 µV/h after 20,000 h operation is equivalent to 100–200 mV cell increase). When analyzing the degradation of a PEM water electrolysis cells, there are various considerations that need to be taken into account and different aspects to bear in mind:

- Is the cell at BoL an ideal cell that satisfies the assumption of homogeneity?
- Because ageing strongly depends on operating conditions, a degradation mechanism prevailing in given conditions might differ from one prevailing in other operating conditions.
- In a stack, which cell component is ageing the faster?
- Which microscopic phenomena play a role and what are their physical meanings?
- How can degradation be mitigated in order to increase lifetime?

3.2 EXPERIMENTAL TOOLS FOR INVESTIGATING DEGRADATION PROCESSES AND MECHANISMS

3.2.1 The Measurement Cell

To avoid endless discussions on pathological cases, it is preferable to use state-of-art measurement cells. Fig. 3.2A provides an example of a typical laboratory monocell test bench used for atmospheric experiments. Fig. 3.2B shows an example of a laboratory monocell used for pressurized experiments (Elamet et al., 2013). The design of these cells needs to be mechanically optimized to remove mechanical heterogeneities and ensure that experimental IV curves provide real information

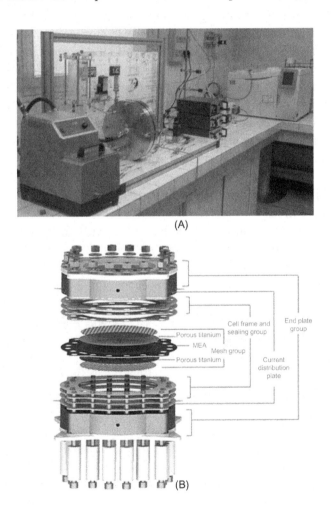

Figure 3.2 (A) Photograph of a laboratory PEM water electrolysis monocell for pressurized experiments. (B) 3D drawing of a laboratory PEM water electrolysis monocell for atmospheric experiments.

on the level of electrochemical performance. Otherwise, only averaged values of heterogeneous current lines are measured.

In PEM water electrolysis, the electrolyte is confined within the thin polymeric membrane of homogeneous thickness. In order to obtain homogeneous current lines across the cell (constant current density over the surface of the cell), the cell should be designed in such a way that homogeneous compression fields are obtained. This is a prerequisite before degradation processes can be analyzed, in order to avoid experimental bias. This is usually obtained by using thick stainless steel external flanges, which also contribute to thermal homogeneity. Fig. 3.3A shows the mapping of compression forces. In Fig. 3.3A, the distribution is homogeneous over the entire surface area. The pattern is due to the PTL. In Fig. 3.3B, there is a strong (and inappropriate) radial distribution of compression forces, with a central section that experiences four to five times less pressure than the periphery. During operation, the current flows mainly across the compressed ring at the periphery of the cell and the experimental IV curve provides only information on the mean cell behavior. Ageing of such a cell is heterogeneous, and data analysis might lead to erroneous conclusions.

3.2.2 In Situ Measurement Techniques
3.2.2.1 I−V Curves
When adequately measured, I−V curves provide key electrochemical information on PEM water electrolysis cells. To avoid experimental bias, measurements must be made under real isothermal conditions. The use of two thermostats, one on each of the anode and cathode

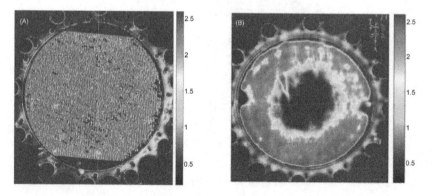

Figure 3.3 (A) Homogeneous distribution of compression forces over the active area. (B) Heterogeneous distribution of compression forces over the active area.

loops, is preferable, especially when cells with large surface areas are tested. In addition, it is necessary to be careful when measurements are recorded at elevated current density because internal dissipation processes in both the membrane (ohmic dissipation) and the catalyst layers (CLs) (charge transfer dissipation) may lead to significant local temperature gradients. Considering the fact that the useful electrochemical information is found in the activation region, it is preferable to restrict the current density (j) range of measurement to the first hundred mA/cm^2, and to go to higher values only to check the absence of possible mass transport limitations. It might also be necessary to measure the IV curve from a j_{max} to j_{min} value instead of j_{min} to j_{max} after thermal stabilization at j_{max}. When isothermal conditions are satisfied, it might be useful to record the I–V curve in both directions (increasing and decreasing j) because hysteresis might reveal mass transport limitations.

Experimental IV curves can be fitted using a simple model, Eq. (3.1), in which the different cell voltage contributions are added (Millet, 2015, 2016):

$$U_{cell}(T,P) = E(T,P) + j \sum_{k=1}^{n} R_k + \eta_{H_2}(T,P) + \eta_{O_2}(T,P) \qquad (3.1)$$

In Eq. 3.1, $E(T,P) = \Delta G(T,P)/2\,F$ is the thermodynamic water-splitting voltage at operating T,P conditions; the R_k are the resistances of the different cell components (including interfacial resistances); η_{O_2} and η_{H_2} are the oxygen evolution reaction (OER) and hydrogen evolution reaction (HER) overpotentials at (T,P), calculated from the Butler–Volmer (BV) equation. Away from equilibrium conditions, charge transfer overpotentials are approximated by the following:

$$\eta_{O_2} \approx \frac{R_{PG}T}{\overleftarrow{\alpha_a}F} Ln\left(\frac{i^a}{i_0^a r_f^a}\right) \qquad \eta_{H_2} \approx \frac{R_{PG}T}{\overrightarrow{\alpha_c}F} Ln\left(\frac{-i^c}{i_c^0 r_f^c}\right) \qquad (3.2)$$

Eq. (3.1) is very general, and independent of cell design. It is simple, easy to use, and contains most of the physics of the water electrolysis reaction. However, it is based on a set of simplifying assumptions that the reader should bear in mind.

- Homogeneity: this is a 1D equation; each cell component is assumed to have homogeneous and isotropic physical properties; temperature and pressure are homogeneous throughout the cell; equipotential lines are parallel to each over the entire cell thickness and current density is constant everywhere in the cell, on any scale (macroscopic and microscopic); such conditions are usually satisfied when the distribution of compression forces is homogeneous (see Fig. 3.3 and related discussion).

- Mass transport limitations, if any, are neglected.
- The CLs are homogeneous monolayers (their thickness is equal to zero) of catalysts placed on each side of the membrane. All active sites (i_0 times r_f) are active at any current density.

Fig. 3.4 shows two typical polarization curves (plots of the cell voltage U_{cell} as a function of operating current density j, at constant operating temperature T and pressure P) recorded at $T = 80°C$ and $P = 1$ bar. The top curve was measured on a Pt/Nafion117/Pt MEA and the second one was measured on an IrO_2/Nafion117/Pt MEA. A reference electrode (Millet et al., 1990) was used to measure, separately, the anode and cathode charge transfer overpotentials as well as the cell voltage drop across the membrane (non-MEA resistance contributions are included in the ohmic drop term). These experimental curves were fitted using Eq. (3.1). Fit parameters and best fit values are listed in Table 3.1. The reference exchange current density (j_0) for the OER measured on a smooth ($r_f = 1$) iridium electrode (j_0^a) was taken from Damjanovic et al. (1966). The reference exchange current

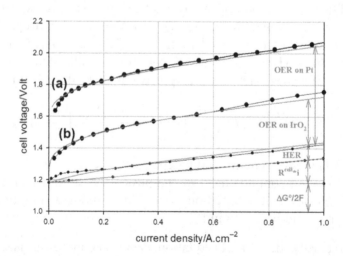

Figure 3.4 Cell voltage breakdown analysis for a conventional PEM water electrolysis cell (see text for details). (●) experimental data; (−) best fits using Eq. (3.1). (a) anode: 2.0 mg/cm² Pt ; cathode: 1.1 mg/cm² Pt. T = 80° C. (b) anode: 2.7 mg/cm² IrO₂ ; cathode: 1.1 mg/cm² Pt.

Table 3.1 Parameters Used in Eq. (3.1) to Fit Experimental Polarization Curves of Fig. 3.4. $T = 80°C$, $\beta^a = \beta^c = 0.5$, $E° = 1.18$ V (roughness factors r_f have no dimension)

	$R_{cell} = \sum_{k=1}^{n} R_k (m\Omega/cm^2)$	j_0^a (A/cm²)	r_f^a	j_0^c (A/cm²)	r_f^c
Pt anode	160	1×10^{-9}	5	1×10^{-3}	40
IrO₂ anode	160	1×10^{-6}	75	1×10^{-3}	40

density for the HER on a smooth platinum electrode (j_0^c) was taken from Bockris and Reddy (1970). Roughness factors of anode and cathode (r_f^a and r_f^c) were adjusted during the fit. In both cases, the HER overpotential is small compared to the other terms of the cell voltage.

When a reference electrode is implemented in the cell, it is possible to measure, separately, the different terms of the cell voltage, in particular the cell resistance and the charge transfer contributions.

3.2.2.2 Cyclic Voltammetry

The two CLs of the PEM water electrolysis cell can be electrochemically probed in situ by cyclic voltammetry (CV). Taking advantage of the facts that the cathode of the PEM water electrolysis cell is a reversible hydrogen electrode (RHE) and that the HER overpotential is small (especially in the activation area of the IV curve, typically for current densities <0.1 A/cm²), the CL at the anode can be probed selectively by CV, using the cathode as reference and counter electrodes. Fig. 3.5A shows typical cyclic voltammograms measured in situ at the anode of PEM water electrolysis cells, at a constant scan rate but for different IrO₂ loadings (Rozain et al., 2016a,b). The normalized charge of these cyclic voltammograms is directly proportional to the IrO₂ loading. This offers the opportunity to probe the concentration of OER sites of interest (those within the CLs that are electronically connected to the cell and available for the OER). It should be noted here that the measured charge depends on the scan rate (Fig. 3.5B). Such dependence is common for oxides. A more detailed analysis reveals that a difference can be made between two populations of electrocatalytic sites: Those available at the surface of the catalyst particles and involved in the OER, especially at elevated current densities, and those deeply embedded within the pores of the catalyst particles (TEM analysis reveals that, in some cases, catalyst particles contain pores or less accessible sites) or not adequately located in the CLs. This is useful

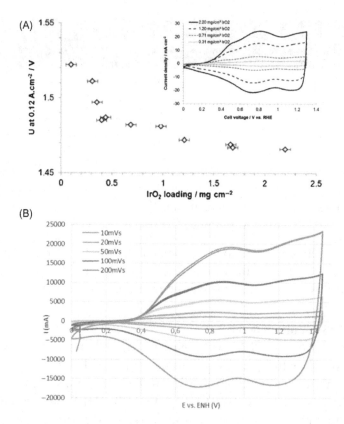

Figure 3.5 (A) Cyclic voltammograms measured on a PEM water electrolysis anode for different anode loadings (@ 20 mV/s) and cell voltage vs loading @ 0.12 A/cm². (B) Cyclic voltammograms measured in situ on a PEM water electrolysis anode at different scan rates (Verdin, 2017).

information, particularly in our efforts to gain a better understanding of how these sites contribute to the overall observed kinetics.

Examining the cathode layer using the same technique is not possible because the anode cannot be used as an internal reference (due to the irreversibility of the water/oxygen redox couple, the electrode potential under constant P_{O_2} is not constant). However, the cathode can be examined using an internal reference electrode, as shown elsewhere (Millet et al., 1993). Fig. 3.6A (Rozain and Millet, 2014) shows examples of Pt cyclic voltammograms in two different electrolytes. In a deaerated sulfuric acid (H_2SO_4) solution, typical cyclic voltammograms measured on a bulky Pt foil clearly show two reversible adsorption/desorption peaks in the H adsorption region (curve a). These two peaks are due to H adsorption/desorption on the two major crystallographic sites found on polycrystalline Pt. This cyclic voltammogram is

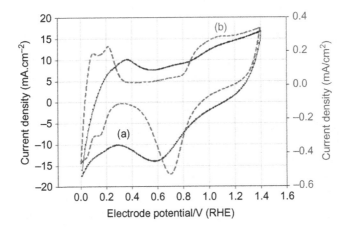

Figure 3.6 Cyclic voltammograms measured on Pt/H⁺ interfaces: (a) Pt foil in H₂SO₄, (b) cathode of a PEM cell (50 mV/s).

the reference case: the curve is symmetrical versus $j = 0$ and the sloppy profile in the H adsorption region is due to the ohmic drop. The reference charge (Trasatti and Petri, 1992) measured on a smooth polycrystalline platinum electrode is $210\ \mu C/cm^2$. In a PEM cell (curve b), the shape can be significantly distorted, depending on how clean the interface is.

On clean surfaces, there is no significant difference between the sulfuric acid and PEM electrolyte environments. The charge required for the surface sorption of one monolayer of atomic hydrogen can be used as a measure of the density of platinum sites available for the HER. This density of sites is proportional to the platinum loading within the cathode. See Fig. 3.7.

3.2.2.3 Interfacial Contact Resistance

In a PEM water electrolysis cell, there are several metal—metal interfaces. The surface roughness of metallic cell components involved in these interfaces is the source of contact resistances, which are usually much larger than the bulk resistance of these metallic cell components. Interfacial contact resistance (ICR) (expressed in $m\Omega/cm^2$) can be experimentally measured using ex situ techniques (Fig. 3.8A). Fig. 3.8B shows some typical ICR values at BoL and EoL when no surface treatment (protective coating) is used. ICR values at BoL and EoL can be significantly different and the impact that such changes may have on the efficiency of the cells can be quite significant.

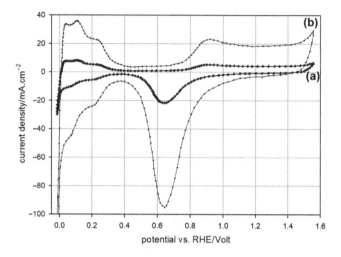

Figure 3.7 Cyclic voltammograms measured on a 2.3 cm² PEM water electrolysis cathode with different Pt loadings: (a) ~0.5 mg/cm², r_f ~ 50; (b) ~1.1 mg/cm², r_f ~220.

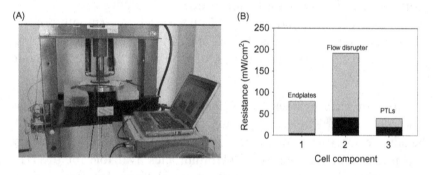

Figure 3.8 (A) Experimental measurement of sample resistance and ICR. (B) Order of magnitude of ICR in a PEM water electrolysis cell at BoL (black) and EoL (gray).

3.2.2.4 Current Distribution

Fig. 3.9A shows a photograph of a segmented bipolar plate used for the in situ measurement of current line and temperature distributions in a PEM water electrolysis cell (INSIDE, 2017). This innovative printed circuit board (PCB) was initially designed and developed for fuel cell applications. It has been adapted for operation in a PEM water electrolysis cell. It can be mounted directly in the PEM water electrolysis cell and used as a bipolar plate with a flow field for water circulation. It can be used for the in situ recording of local current density and temperature during either stationary or transient operation.

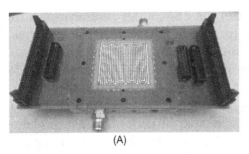 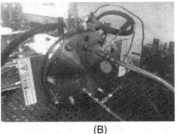

(A)   (B)

Figure 3.9 (A) Segmented bipolar plate for PEM cell with PCB for current density and temperature distribution measurements (INSIDE project, 2017). (B) 250 cm² PEM water electrolysis circular monocell equipped with a PCB (Verdin, 2017).

These somewhat recent developments are commercially available (S++ Simulation Services) and can potentially be used for design qualification, design optimization, and in situ analysis of degradation processes.

3.2.3 Ex Situ Measurement Techniques

All chemical or physical ex situ techniques that can provide information on any cell component are potentially interesting and useful. This is particularly true for the electrocatalysts and polymer electrolytes. Among the most popular techniques, let us mention those that provide information on bulk chemical composition (X-ray diffraction, XRD) or surface chemical composition (X-ray photoelectron spectroscopy, or XPS). New techniques requiring specific and nonroutine equipment are also increasingly being used. These include, e.g., near ambient pressure XPS (NAP-XPS), which provides surface information in an electrolyte environment, and X-ray absorption near edge structure spectroscopy (XANES) and extended X-ray absorption fine structure spectroscopy (EXAFS), available at synchrotron radiation facilities. These analytical techniques provide detailed, and quite useful, information on electrocatalytic sites on the atomic scale, with or without electrolyte and external polarization. Atomic force microscopy (AFM) coupled to scanning electrochemical microscopy (SECM) is also being used. Furthermore, various ex situ electrochemical techniques, such as scanning CV or nonharmonic impedance spectroscopies can be used to determine electrochemical activity in various environments.

Among the different cell components of the PEM water electrolysis cell, the MEA is usually the most scrutinized component because the

MEA is the weakest component of the PEM water electrolysis cell (or, at least, the component that is the most prone to degradation). Postoperation and postmortem analyses are mainly used to investigate the structure of these MEAs, in particular the membrane (looking for pinholes or thinning effects) and the CLs (looking for possible detachment or dissolution of catalyst particles to explain reduced activity). They are the two most probable sources of performance degradation when the PEM cell is operated for long periods of time. Scanning electron microscopy (SEM) and transmission electron microscopy (TEM) often provide useful information.

Metallic cell components are usually characterized by nondestructive surface characterization tools. Titanium (Ti), for instance, which is extensively used in medicine and biomedical applications, is characterized using variety of techniques, such as XPS or NAP-XPS, Auger electron spectroscopy (AES) and scanning Auger microscopy (SAM), secondary ion mass spectroscopy (SIMS), Rutherford backscattering spectrometry (RBS), and nuclear reaction analysis (NRA) (Lausmaa et al., 1990).

3.2.4 Accelerated Stress Test Protocols

Accelerated stress test (AST) protocols can be used to accelerate degradation processes and reduce the duration of experiments. The purpose of such approaches is to design specific tests that can simulate the lifetime of the cell to within days or even weeks, instead of years. The interest of such tests is that they provide general information on degradation phenomena (mechanisms and rate of degradation). However, when such tests are not well designed, this information does not necessarily inform on what will happen in real operating conditions.

Stress factors implemented in ASTs to accelerate ageing and identify underperforming cell components or those prone to fast ageing include the following:

- Transient operating power sources (the selection of the profile of interest will depend on the target application)—power profiles are usually designed using either transient current density or transient cell voltage profiles
- Nonisothermal operation
- Maximum anode potential

- Membrane swelling (due to temperature changes and also water flow)
- Thermal gradients and differential swelling in CLs
- Rest potential when the cell is switched off or in an idle state

The simplest AST is a continuous (stationary) chrono-potentiometric (CP) experiment at different T, P, and j operating conditions. The two main stress factors are operating current density and temperature. Other ASTs are based on nonstationary CP or CA operating conditions. Fig. 3.10 shows two examples of such nonstationary ASTs of interest for PEM water electrolysis cells or stacks. In both cases, this is a succession of CA steps of different magnitude and frequency. Fig. 3.10A shows a succession of galvanostatic pulses. Such profiles are used to simulate on/off experiments such as those required for grid services (see Chapter 2, Key Performance Indicators). Fig. 3.10B presents a periodic staircase profile. Such profiles of parabolic shape can be used to simulate photovoltaic (PV) power sources. The hybridation of the two profiles can provide a more realistic simulation of a PV source. A major difference between the two tests is that, during the first one, the cell remains always polarized. Hence, specific phenomena occurring under open-circuit conditions are not explored. Another difference is that, during the first test, resting periods may induce thermal fluctuations if the cell is not thermostated. This is usually done intentionally. Another difference is that, in the second test, the current density is high and the anode is likely to operate at OER voltages >2.0 V (a threshold value that usually triggers interface corrosion of titanium cell components when they are not surface treated). Thermal changes (due to fluctuating current density or overcapacity operation or water starvation) and

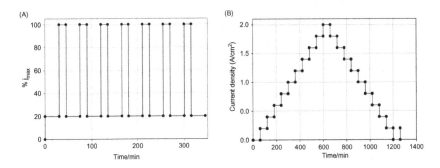

Figure 3.10 (A) Example of pulsed AST protocol with period peak currents and galvanostatic resting periods. (B) Example of staircase galvanostatic AST protocol.

related water swelling effects are specific and significant stress factors for the membrane and the catalytic layers.

In reality, the design of ASTs is not done randomly or based on imagination. The AST should be designed to accelerate the effects of the main features of the target application. Ultimately, the actual degradation information should be obtained when the cell/stack is operated using the same operating conditions as in real life. The power source is the key factor that indicates what might happen during operation. If the cell is designed for operation with a strongly fluctuating power source, such as solar-PV, or wind, for specific applications, such as for stationary or grid services, then such load profiles should be taken into consideration in the tests.

3.3 DEGRADATION MECHANISMS

3.3.1 Overview

There are some pathological cases that are due to the fact that nonconventional materials are used as cell components or that the cell design is inappropriate (e.g., the cell contains components prone to chemical or electrochemical corrosion). The following factors also have deleterious effects:

• Nonhomogeneous field of clamping pressure
• Use of insufficiently deionized water
• Inappropriate MEA manufacturing

Such pathological cases are not considered here. When the cell is appropriately designed, a quick look at Fig. 3.1 tells us which factors induce degradation:

• Increasing bulk resistances (ionic or electronic)
• Increasing interfacial resistances (electronic)
• Increasing charge transfer overvoltages

The main degradation mechanisms are as follows:

• Platinum oxidation and dissolution at the cathode at open-circuit conditions after polarization (platinum particles are found at different places within the membrane)
• Hydrogen peroxide formation within the membrane at specific locations where H_2 and O_2 reactants + Pt particles (catalyst) are

present and local potential is appropriate (H_2O_2 has a deleterious effect on PFSA chains)
- Loss of catalytic sites in either anode or cathode CLs due to inappropriate CL engineering or thermal gradients, and local membrane swelling differences or local heterogeneous distributions of current lines
- Oxidation at metal−metal interfaces
- Corrosion of carbonaceous PTLs whenever used

The main consequences of degradation are as follows:

- Chemical or physical alteration of the membrane, leading to either an increase in ionic resistivity or a chemical oxidation that leads to membrane thinning and, in a paradoxical way, to improved cell performance, yet higher cross-permeation
- Loss of energy efficiency
- Loss of coulombic efficiency
- Loss of gas purity
- Increasing safety issues and risk of irreversible breakdown

3.3.2 Membrane Degradation
3.3.2.1 Reversible Modification of Membrane Bulk Ionic Conductivity
For good (efficient and durable) operation, a PEM water electrolysis cell requires the use of highly purified (ionically and biologically) water. During operation, there are two main constraints that need to be satisfied: (i) the need to supply the stack with water of adequate purity and (ii) the need to maintain that purity during operation. Hence, tap water should be biologically and ionically purified prior to its introduction into the electrolysis module. At ambient conditions the ionic product of water is 1×10^{-14}. The self-ionization of water induces a residual conductivity equivalent to a resistivity of ~ 18 MΩ/cm. There is no need to reach such values to feed a PEM water electrolysis cell. A resistivity of 1 MΩ/cm (equivalent to a residual conductivity of 1 μS/cm[1]) is sufficient.

There are several factors that tend to degrade water purity (purity is usually measured online using a conductimeter):

- Biological contamination by air contact (requires a UV lamp)
- Metallic cations that are released by corrosion of steel piping in the circuitry or titanim cell components
- Membrane ageing that tend to release fluorine ions

If water used in the process has not the appropriate level of ionic/organic purity, then the impurities tend to accumulate within the membranes and membrane performances tend to decrease. Cations are incorporated into the PFSA membrane via an ion-exchange process with internal hydrated protons. Concentration levels that drive the process are thermodynamically set by sorption isotherms (the quantitative relationship between external and internal ion concentration). The process is reversible; however, cleaning is not an easy task. In most cases it is necessary to dismantle the cells—a practice that has significant drawbacks (because CLs strongly embedded in the first layers of the PTLs tend to peel off) and one that cannot be repeated, except when CLs can be adequately regenerated.

An example of such contamination is provided by the EoL chemical analysis of the membrane of a Pt/Nafion117/IrO$_2$ MEA that was operated under stationary conditions (chrono-potentiometry at $+2.0$ V) for 5000 h. At the end of the experiment, several metallic ions where found in the membrane (mainly iron, cobalt, nickel). These ions were all released during the experiment by surface corrosion of the stainless steel tubing through which deionized water was pumped. Fe^{3+} ions accounted for 24% of the total SO$_3^-$ sites, Ni^{2+} for 2.5% of the total SO$_3^-$ sites, and Cr^{3+} for 0.7%. Gradual corrosion of the tubing was found to increase the concentration of these ions within the water under circulation. Donnan exclusion can prevent incorporation as long as the ion concentration remains low, but at higher concentration values, these cationic species can penetrate the membrane via an exchange process with internal protons. As a result of proton release, the acidity of the circulating water flow, and hence the corrosion rate, tend to increase. This membrane was regenerated by immersion for 2 h in a boiling aqueous solution of sulfuric acid followed by immersion for 2 h in boiling deionized water to remove traces of remaining sulfuric acid.

The residual conductivity of water after treatment and within the circulation loops must be kept below this threshold value of 1 μS/cm[1] (resistivity > 1 MΩ/cm). In commercial systems, this can be achieved by using online conductivity monitoring and ion-exchange resin beds that can adsorb foreign ions and other impurities, and maintain the residual conductivity at appropriate levels. It should be noted here that the use of online ion-exchange resins can reduce the operating temperature to a maximal value of 60°C only. The need for highly deionized water is undoubtly a disadvantage here compared, e.g., to alkaline water electrolysis; it adds extra costs to the process. Nonetheless, the situation can be improved by using nonplatinum HER electrocatalysts.

3.3.2.2 Irreversible Chemical Degradation of Membrane and Thinning

PFSA materials used as solid polymer electrolyte membranes in PEM water electrolysis cells are prone to chemical attack and corrosion. PFSA materials are characterized by their equivalent weight (EW) (this is a measure of the weight of polymer, in gram, that corresponds to one equivalent of electric charge). Membranes of different EW and thickness are commercially available. Polymer side chains are prone to chemical corrosion via various microscopic mechanisms that depend on operating conditions (e.g., the local potential and chemical composition inside the membrane). A reduction of the concentration of charge carriers is equivalent to an increase in EW. As a result, the electrical resistance of the membrane will increase.

The most spectacular consequence of such degradation is a gradual reduction in the membrane thickness (membrane thinning) and/or pinhole formation (Stucki et al., 1998). A quantitative example of membrane thinning during water electrolysis operation is provided in Fig. 3.11 (Fouda-Onana, 2016). Fig. 3.11(A) shows the details of the AST protocol used during the evaluation: this is a succession of CA steps between 0 and 1.6 A/cm^2. A cross-section of a PEM water electrolysis MEA at BoL is shown in Fig. 3.11B, and at EoL in Fig. 3.11C. Postmortem analyses revealed a strong chemical corrosion of the polymer membrane; after 1500 h of AST, the thickness of the membrane had decreased by 30 μm, corresponding to a mean degradation rate of 20 nm/h. Fluorine emission analysis confirmed the chemical attack—a kinetics of the process can be obtained by measuring the fluorine content during the test. The kinetics is impacted by the higher current density value (higher anodic potential at the anode) used in the

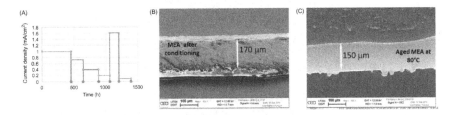

Figure 3.11 (A) AST protocol used in the experiments. (B) SEM micrograph of MEA at BoL. (C) SEM micrograph of MEA at EoL.

AST, but also by the resting periods under open-circuit conditions. Chemical mechanisms leading to membrane degradation have been extensively studied in PEM fuel cell technology (Luan and Zhang, 2012). A specific mechanism involves the formation of hydrogen peroxide, H_2O_2 (in a PEM water electrolysis cell, hydrogen peroxide forms as a result of H_2/O_2 cross-permeation and chemical recombination on specific surfaces at specific potentials). Various chemically active radicals then form and can attack some weak chemical bonds (e.g., some ending $-COOH$ groups founds in PFSA materials), leading to the formation of carbon dioxide (CO_2) and hydrofluoric acid (HF). In addition, the release of HF can have significant deleterious effects (e.g., corrosion of titanium cell components).

Different mitigation measures can be implemented. For example, membrane chemistry and engineering is crucial: fluorination of carboxylate end-chain groups and minimization of ether cleavage by using short-side chain structure. Another option is to introduce radical scavengers within the membrane/CLs) but developments for PEM water electrolysis cells are still scarce.

3.3.2.3 Membrane Perforation
The combination of various factors can accelerate membrane degradation. Ultimately, this can lead to membrane perforation. An example is provided in Fig. 3.12. The catalytic recombination of hydrogen and oxygen stored in the electrolysis compartments can have quite dramatic consequences, and ultimately lead to the destruction of the PEM

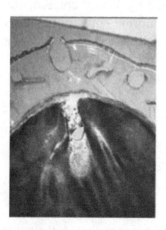

Figure 3.12 Photograph of a perforated MEA (the white spots contain mainly PFSA degradation products).

cell/stack (Millet et al, 2012). An appropriate management of safety issues requires the continuous monitoring of hydrogen/oxygen purity and individual cell voltages.

3.3.3 Catalyst Degradation

Catalyst degradation has multiple origins. Depending upon operating conditions, one mechanism or another can become more or less critical. As can be seen from Eq. (3.1), there are two main causes of catalyst degradation and loss of electrochemical activity in the CLs: (i) loss of intrinsic activity (via j_0) and (ii) loss of active site (via r_f). Problems of the first category are usually due to surface contamination of catalyst particles, especially at the cathode of the cell because Pt, which is very active, is also highly prone to surface contamination by Under Potential Deposition (UPD) or bulky deposition of metallic contaminants. Problems of the second category have quite diverse origins. They occur both on the anode and cathode side of the cell. To monitor such ageing processes, it is convenient to use in situ electrochemical tools such as those described in Section 3.2.

3.3.3.1 Loss of Intrinsic Activity

These different types of degradation mechanisms are due to chemical composition changes at the surface of catalyst particles. This may be the result of surface poisoning or surface segregation/corrosion when alloys or solid solutions are used, at a constant concentration of catalytic sites (e.g., when IrO_2/RuO_2 solid solutions are used at the anode for the OER). An example of such reversible contamination is shown in Fig. 3.13 (Millet et al., 1990). This figure shows IV curves recorded on an IrO_2/Nafion117/Pt° MEA after equilibration in aqueous $NiCl_2$ solutions of increasing concentration. Donnan equilibrium leads to an increasing Ni^{2+} concentration within the membrane. As the nickel concentration increases, the IV curve is shifted toward high cell voltage values but the slope at high current density remains unchanged. The interpretation of these results is that once nickel ions are inside the membrane, they react to the electric field and migrate to the cathode where they are electrochemically reduced to form under potential or bulk deposits at the surface of platinum catalyst particles. As a result, the HER overpotential increases gradually as less platinum sites are left uncovered, while the membrane resistance remains unchanged. The position of the wave on the I-V curve is related to the fraction of uncovered Pt sites. The process is reversible. After reequilibration in

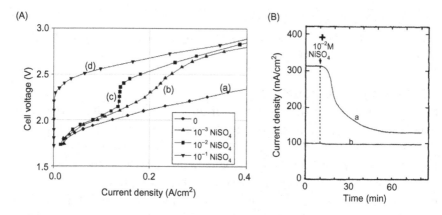

Figure 3.13 (A) IV curves measured on an MEA after equilibration with aqueous $NiCl_2$ solutions of increasing concentration (Millet et al. 1993). (B) Chrono-amperometric plots measured on a Pt/Nafion117/Pt MEA (S = 5.3 cm²). Constant cell voltage (a) 2.35 V and (b) 2.05 V (Andalfatto et al., 1994).

an aqueous acidic solution, nickel ions are removed and the IV curve returns to its initial position. Similar results are found during CA experiments. This is illustrated in Fig. 3.13B, where the PEM water electrolysis cell is operated at two different operating current densities: 100 and 300 mA/cm². The larger the operating current density the lower the potential of the HER cathode. At 300 mA/cm², when water circulating in the anode cell compartment is replaced by an aqueous solution of 10^{-2} M $NiCl_2$, the current density decreases. At 100 mA/cm², the impact is almost negligible. These data can be interpreted as those of Fig. 3.13A: There is a critical cathode potential value below which there is a metallic deposition of nickel on platinum that increases the HER overpotential.

3.3.3.2 Loss of Active Sites

Degradation means that the population of active sites tends to decrease during operation. Fig. 3.14A shows the cyclic voltammograms measured on the anode of the PEM water electrolysis at BoL and EoL. A similar plot for the cathode is shown in Fig. 3.14B. The reasons for such reductions in the populations of catalyst sites are catalyst particle losses (transport away the CLs, corrosion/dissolution in the water stream and/or the polymer electrolyte), and catalyst particle sintering (e.g., Pt nanoparticles coated at the surface of carbonaceous substrates are prone to surface mobility and sintering, even at moderate operating temperatures).

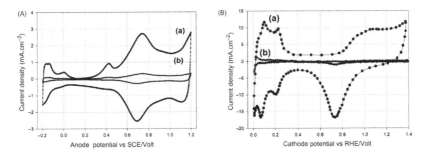

Figure 3.14 (A) CV curves measured on a PEM water electrolysis anode at (a) BoL and (b) EoL. (B) CV curves measured on C/Pt PEM cathodes at (a) BoL and (b) EoL.

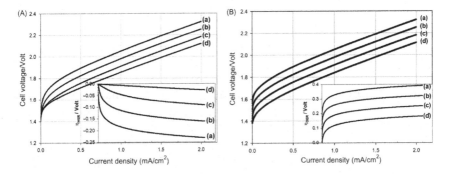

Figure 3.15 (A) IV curves calculated using Eq. (3.1) for different HER roughness factors ($r_f^a = 100$): (a) $r_f^c = 1$, (b) $r_f^c = 10$, (c) $r_f^c = 100$, and (d) $r_f^c = 1000$. (B) IV curves calculated using Eq. (3.1) for different OER roughness factors ($r_f^c = 100$): (a) $r_f^a = 1$, (b) $r_f^a = 10$, (c) $r_f^a = 100$, (d) $r_f^a = 1000$.

The impact of the roughness factor of the CLs is visible in the activation domain of the IV curves, for current density values between 0 and ~ 0.1 A/cm². Fig. 3.15A shows the IV curves calculated using Eq. (3.1) at constant r_f^a for different values of r_f^c (the insert shows the HER overpotential versus the operating current density). Fig. 3.15B shows the IV curves calculated using Eq. (3.1) at constant r_f^c for different values of r_f^a (the insert shows the OER overpotential versus the operating current density). A good indicator of the number of OER sites is found in the activation domain at very low current density values because the exchange current density of the water/oxygen redox couple is 1000 times less than the exchange current density of the water/hydrogen redox couple. When the exponential rise on the cell voltage remains close to the cell voltage axis and expends $>1.5-1.6$ V, this is a good indication that the number of OER sites is insufficient. Furthermore, a large overpotential increase at very low current density

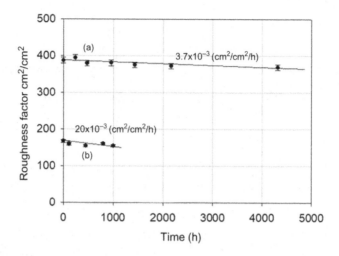

Figure 3.16 Anode roughness factor versus time: (a) 1.5 mg_{IrO2}/cm² at 1.0 A/cm² and (b) 0.9 mg_{IrO2}/cm² at 1.5 A/cm².

is the signature of anode degradation. When the shape of the I–V curve in the activation area changes at higher current density values, this can be more likely ascribed to the cathode.

Fig. 3.16 shows a measure of the roughness factor (r_f) of PEM water electrolysis anodes operating at different stationary current densities. Curve (a) shows the situation for a high loading anode operating at low current density (j). Curve (b) shows the situation for a low loading anode operating at higher current density. Such degradation processes involve quite different mechanisms. The predominance of any of these mechanisms depends largely on the engineering of the CL (materials + manufacturing) and the operating conditions.

3.3.4 Catalyst Layer Degradation
3.3.4.1 Global Heterogeneous Current Distribution
As already discussed, current lines through the PEM cells can be quite heterogeneous. Fig. 3.17 shows a distribution of 0.4 ± 0.1 A/cm² corresponding to $\pm 25\%$ of the mean current density value across the cell. Such an in situ diagnostic tool is very sensitive and very helpful to obtain information on the homogeneity of current line distribution and associated thermal gradients. It is surprising, and unexpected, to find that even on cells with small sections, the distributions can be quite heterogeneous. This is a very good indicator of the appropriateness, or the lack thereof, of the cell design. Of course, ageing is also quite

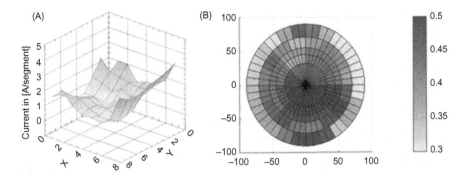

Figure 3.17 (A) Current distribution measured on a 250 cm² PEM water electrolysis cell during stationary operation (raw data). (B) Current distribution measured on a 250 cm² PEM water electrolysis cell during stationary operation (post-treated data).

heterogeneous; however, the analysis made under the assumption that the current lines are evenly distributed can be totally misleading.

3.3.4.2 CL Structure Modification

As discussed in Chapter 3, Fundamentals of Water Electrolysis, Volume 1, CLs are critical elements of the PEM cell. For optimal operation, this is a region where at least three/four percolating phases must coexist (for the transport of electrons, ions, liquid water and gases). Triple/quadruple point populations on each side of the membrane dictate efficiency and should be maximized. Compared to PEM Fuel Cells, no specific measure is implemented in PEM water electrolysis cells to promote gas transport (e.g., pore promoters are not used; also, the EW of the CLs is not different from the EW of the bulk membrane because gas solubility in these layers is not a critical issue in terms of mass transport). Regarding degradation, several external factors, such as thermal cycling, local water drying, and heterogeneous distribution of current lines, are suspected to impact the structure and stability of the CLs, and hence their performance.

As for any other cell component, the rate of degradation depends upon specific operating conditions (applications of the water electrolysis cell) but even under stationary operating conditions, there are problems. The local scale (micrometer scale) distribution of current lines, at the interface between the PTL and the CL, can be quite heterogeneous. This is because the PTL is a porous structure and the distance between contact points with the CLs can be significantly large, depending on the shape and geometry of Ti particles and the resulting PTL porosity. The resulting local thermal gradients are responsible for differential water swelling and water content, and in turn inducing a stress that can contribute to CL ageing. Fig. 3.18A shows a SEM micrograph of a Ti-PTL made of sintered spherical Ti particles. Fig. 3.18B shows an SEM top-view micrograph of the anode CL after contact with the PTL of Fig. 3.18A (Millet et al., 2011). The craters are due to the compression of the first row of spherical Ti particles onto the mechanically soft IrO_2 + Nafion CL. Cracks observed around these regular craters indicate where the gaseous oxygen flow is collected during electrolysis.

The PTL can be considered as a fragmented electrode. The surface of direct contact between the PTL and the CL and the distance between contact points depend on the geometry of the PTL used for the experiment. Current distribution at such nonequipotential interfaces can be measured and can be modeled. Fig. 3.19A shows an example of such a

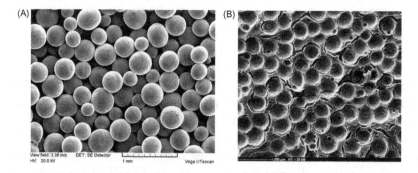

Figure 3.18 (A) SEM micrograph of a Ti-PTL made of sintered spherical Ti particles. (B) SEM top-view micrograph of the anode CL after contact with the PTL of Fig. 3.18A.

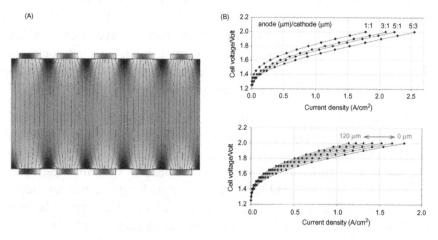

Figure 3.19 (A) Distribution of model current lines in a PEM water electrolysis cell. (B) Calculated IV curves: (top) for various anode/cathode thicknesses; (bottom) for different distances between contact points.

1D model (Verdin et al., 2017). Current lines are calculated by finite element analysis (FEA). Fig. 3.19B shows the calculated IV curves for various anode/cathode thicknesses and for various distances between contact points. The overall shape of the I−V curve remains unchanged but water dissociation is more or less efficient when both parameters are changed. The impact of the PTL geometry is similar: The cell resistance tends to increase when the distance between contact points increases. At the cathode, where the HER overvoltage is small, hot points are more probable to form because the charge overvoltage cannot contribute to local current rehomogenization within the CL.

3.3.5 Porous Current Collector Degradation

PTLs, also called porous current collectors (PCCs), are used in the cell for the dual purpose of carrying electric current between bipolar plates and CLs, and transporting fluids (reactant and reaction products) to/ from CLs. There are concerns about hydrogen embrittlement and/or HF attack (as discussed earlier) in/of titanium plates (bipolar and PTLs) and loss of bulk mechanical properties. However, irrespective of their elemental composition and independent of the cell design, metallic cell components of the PEM water electrolysis cell are more sensitive to surface/interface degradation (*i.e.* ICR resistance increases) than bulk degradation, as discussed earlier. Fig. 3.20A shows a Pourbaix diagram of titanium and a broad domain of potential−pH operating conditions found in a PEM water electrolysis cell. Titanium cell components are prone to surface passivation, which explains the interest of this material for such applications. The price to pay, however, is the formation of nonelectronic-conductor surface oxides that are responsible for large ICR. Fig. 3.20B shows XPS diagrams measured at the Ti-2p energy level on a titanium PTL. The different spectra were recorded after surface abrasion of nanometer-thick layers. Results confirm that titanium PTLs (as well as bipolar plates) are surface protected against corrosion/dissolution by the presence of an extrathin passivating TiO_2 layer located at the surface. The surface layer is rapidly etched, uncovering the contribution of bulk metallic titanium. Surface oxidation with air is fast, but the surface layer remains extremely thin because of the lack of mobility of oxygen ions at ambient conditions. Surface layers tend to grow thicker when anodically polarized.

Fig. 3.21 shows a visual comparison between BoL and after 192 h operation (Georg et al., 2016). The color of the titanium cell

(A)

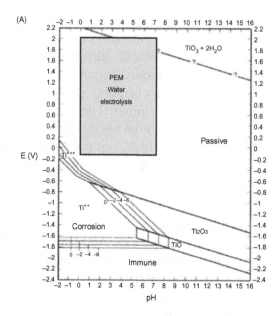

(B)

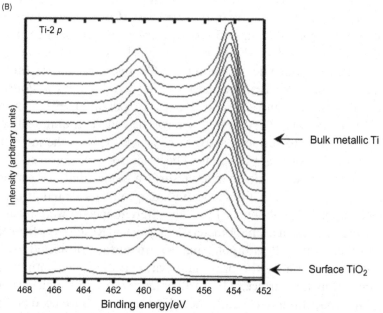

Figure 3.20 (A) Pourbaix diagram for titanium showing the potential−pH range of application for PEM cells. (B) XPS (Ti-2p, 452−468 eV) analysis of the surface of Ti-PTLs.

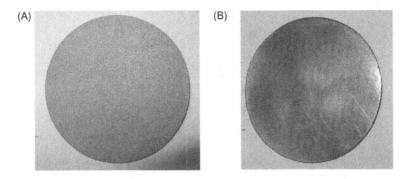

Figure 3.21 (A) Porous current collectors at BoL. (B) Porous current collectors after 192 h operation.

components tends to change during operation. The color is related to the thickness of the surface oxide layer. The yellowish color is consistent with a 20 nm thick surface layer. As a consequence, the contact resistance increases from ca. 1 to 25 mΩ/cm^2. The conduction mechanism across such a thin surface oxide is not yet fully established. However, this surface oxide layer is detrimental to the good operation of the PEM water electrolysis cell, and surface treatments are required to avoid large contact resistances. Many surface treatments have been described in the literature, including platinum sputtering or the formation of surface titanium nitrides/carbides (Millet, 2008).

3.3.6 Bipolar Plate Degradation

Bipolar plates, with or without flow fields, are usually bulk metallic cell components. They are mainly prone to surface modification (corrosion vs oxidation). They are also prone to bulk modifications (hydrogen embrittlement in titanium-based materials). In the short term, this usually leads to increasing electronic resistivity and contact resistances, and also to uneven current distributions. In the longer term, this might also lead to mechanical failure. Poorly performing bipolar plates tend to increase the slope of the IV curves, as any other metallic cell components. Fig. 3.22 shows that massive titanium bipolar plates are prone to surface oxidation on each side of the cell (Georg et al., 2016). This is an indication that contact with deionized water is the predominant factor that causes this oxide layer to grow.

It should be noted here that the term "bipolar plate", used to name the bulky metallic plates used as cell separators in PEM water electrolysis cells, is not totally appropriate, because no electrochemical

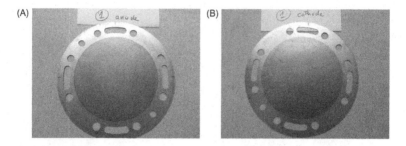

Figure 3.22 (A) Anode side of titanium bipolar plate after 300 h polarization. (B) Cathode side of titanium bipolar plate after 300 h polarization.

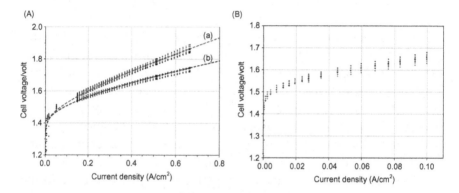

Figure 3.23 (A) IV curves (a) after and (b) before long-term storage (ohmic domain). T = 80°C. (B) IV curves after (black +) and before (red +) long-term storage (activation domain). T = 30°C.

reaction occurs on either side. The electric potential is almost the same on each side. The difference of only a few millivolts (mV), in the worst cases, arises from surface voltage drops on nontreated plates.

3.4 EXAMPLES OF AGEING

3.4.1 Degradation Due to Long-Term Storage of a Proton-Exchange Membrane Stack

The following data were measured on a 12-cell cell stack (250 cm^2 each) that was stored in water for almost 10 years. Fig. 3.23A shows the I–V curves measured at 80°C at BoL and EoL; they only differ by the slope at large current density, which is equal to the internal cell resistance. Fig. 3.23B shows the I–V curves measured at 30°C at BoL and EoL in the activation region ($0 < j < 0.1$ A/cm^2); there is no significant difference between the different curves. The shape of the IV curves in the activation region at low current density reveals that the electrochemical activity of these cells was not significantly affected by this long-term period of inactivity. This was confirmed by the measurement of the anode CVs of individual cells, which were quite similar. However, the steeper slopes from the curves measured after 10 years tell that the cell resistances had significantly increased. A measurement of these slopes at different operating temperature reveals that the membrane resistance remains almost unchanged. It can be concluded that titanium cell components are significantly more oxidized at EoL compared to at BoL, and that the main effect of long-term storage in water is on the titanium bipolar plates, flow disrupters, and PTLs.

3.4.2 Degradation During Stationary Operation
3.4.2.1 Laboratory Cell Degradation
Cell ageing can be followed by either recording IV curves from time to time during a long-term test or via CA or CP experiments (see Section 3.1 to see how this can be done). Fig. 3.24A shows two examples of degradation measured on laboratory cells. The first cell was a $2.3 \, cm^2$ Pt/Nafion117/Pt cell tested for 20,000 h under CP conditions with periodical cleaning. A mean cell voltage drift of 16 μV was measured over that long period of time. Despite the use of Pt at the anode for the OER and the large cell voltage, it is very interesting to note that the cell could be operated almost continuously for such a long period of time. Results obtained for an IrO_2/Nafion117/Pt MEA with a similar surface are plotted for comparison. The degradation rate measured over a shorter period has a similar value to that mentioned earlier (Pt/Nafion117/Pt).

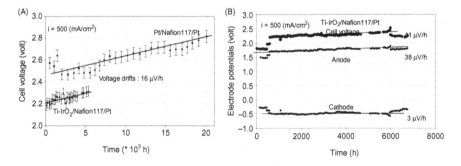

Figure 3.24 (A) Cell voltage versus time of (a) Pt/Nafion117/Pt and (b) IrO_2/Nafion117/ Pt cells at a constant current density of 0.5 A/cm² (Millet et al., 1989). (B) Cell voltage, H_2 and O_2 overvoltages measured on a PEM water electrolysis cell operating at a constant current density of 0.5 A/cm² (Millet et al., 1996).

Using a reference electrode, it is possible to measure, separately, anodic and cathodic potential contributions to the overall cell voltage. Fig. 3.24B shows such a case, where the following are recorded simultaneously: (i) the total cell voltage that tends to increase at a rate of 41 μV/h; (ii) the anode overpotential (including the contribution of half of the membrane ionic resistance) that tends to increase at a rate of 38 μV/h; and (iii) the cathode overpotential (including the contribution of half of the membrane ionic resistance) that tends to increase at a rate of 3 μV/h.

3.4.2.2 Stack Degradation

The main market for PEM fuel cells is the hydrogen mobility sector. Lifetimes of approximately 5000 h are required. The load profile has been standardized—this is the usual profile used for PEM fuel cells (Tsotridis et al., 2015). PEM water electrolyzers are used for hydrogen production and much longer lifetimes are required to reduce current hydrogen costs to a more competitive price.

Fig. 3.25 shows an example of long-term stack durability. During the first thousand hours of operation, the degradation rate is high (up to 30 μV/h in this case). Then gradually it tends to decrease, to a minimum value of 4 μV/h in this case. This 4 μV/h decay corresponds to a 5-year lifetime (160 mV increase of the cell voltage). Projections

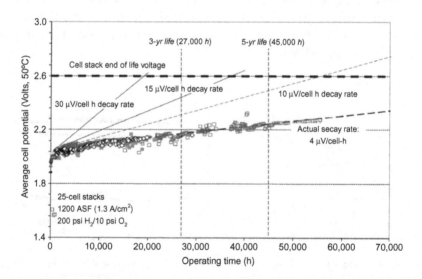

Figure 3.25 Long-term stack durability data (Renner et al., 2015).

indicate that suitable design and water quality should make the technology operational for up to 10 years (80,000 h).

3.4.3 Degradation During Transient Operation

As mentioned earlier, the load profile and operating temperature conditions act as additional stressors. Fig. 3.26 shows the results obtained when a PV profile is used as power source for the PEM water

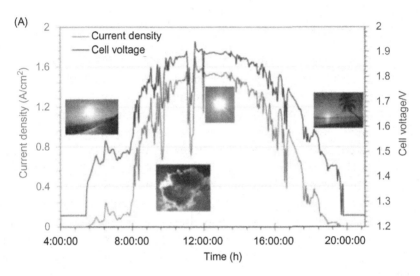

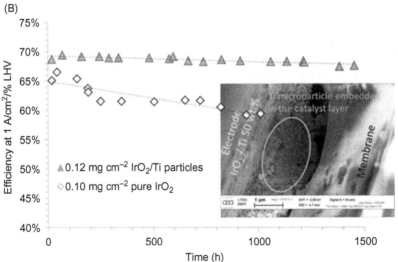

Figure 3.26 (A) Current density versus time of a solar PV panel. (B) Efficiency loss at 1 A/cm².

electrolysis cell (Rozain et al., 2016a,b). The cell efficiency, based on the LHV (lower heating value) of the combustion of hydrogen in oxygen, calculated at 1 A/cm for two different MEAs is plotted as a function of time. The diamond-shaped symbols show the situation for an IrO_2 anode (2% efficiency loss after 1,500h of operation) and the triangles show the situation for an IrO_2 anode containing additional Ti particles used to improve the distribution of current lines within the CL (8% efficiency loss after 1,500h of operation).

3.5 CONCLUSIONS

"Real" PEM water electrolysis cells and stack are not "ideal" ones. It should be noted that a high BoL level of performance does not necessarily mean a high durability. Performances usually tend to decrease with time. This is a multifactor problem, as each cell component has its own weaknesses and limitations. In addition, the ageing rate of these different cell components is strongly affected by the operating conditions. In other words, cell components need to be customized, depending upon the specific operating conditions for which the equipment is used. Technology manufacturers are usually reluctant to provide details about their technology but, as the market is growing, customized designs for the main modes of operation (stationary, grid services, or full transient for PV and wind applications) are now emerging.

REFERENCES

Andolfatto, F., Durand, R., Michas, A., Millet, P., Stevens, P., 1994. Solid polymer electrolyte water electrolysis: Electrocatalysis and long-term stability. Int. J. Hydrogen Energy 19, 421–427.

Bockris, J.O.'M., Reddy, A.K.N., 1970. Comprehensive Treatise of Electrochemistry. Plenum Press, New York.

Damjanovic, A., Dey, A., Bockris, J.O.'M., 1966. Electrode kinetics of oxygen evolution and dissolution on Rh, Ir, and Pt–Rh alloy electrodes. J. Electrochem. Soc. 13, 739–746.

Elamet, O.F., Acar, M.C., Mat, M.D., Kaplan, Y., 2013. Effects of operating parameters on the performance of a high-pressure proton exchange membrane electrolyzer. Energy Res. 37 (5), 457–467.

Fouda-Onana, F., 2016. AST Protocols for PEM water electrolysis: Insight on performances and components degradation. 2nd Int. Workshop on Durability and Degradation Issues in PEM Electrolysis Cells and their Components. 2016 February 17; Freiburg, Germany.

Georg, A., Lickert, T., Smolinka, T., Zhang, X., 2016. Corrosion protective coatings for bipolar plates and current collectors in PEM electrolysers. In: Proceedings of the 2nd Int. Workshop on Durability and Degradation Issues in PEM Electrolysis Cells and their Components. 2016 February 17; Freiburg, Germany.

INSIDE, In-situ diagnostics in water electrolyzers, 2017. European Commission, FCH-JU research project, http://inside-project.eu/, coordination I. Biswas.

Lausmaa, J., Kasemo, B., Mattsson, H., Odelius, H., 1990. Multi-technique surface characterization of oxide films on electropolished and anodically oxidized titanium. Appl. Surf. Sci. 45 (3), 189–200.

Luan, Y., Zhang, Y., 2012. Membrane Degradation, PEM Fuel Cell Mode Degradation Analysis. CRC Press, Boca Raton.

Millet, P., 2008. Final report of the GenHyPEM FP6 research project, FCH-JU. European Commission, Brussel, Belgium.

Millet, P., 2015. Degradation processes and failure mechanisms in PEM water electrolyzers. In: Bessarabov, D., Wand, H., Li, H., Zhao, N. (Eds.), PEM Water Electrolysis for Hydrogen Production: Principles and Applications. CRC Press, Taylor & Francis, Boca Raton, pp. 219–242. Chapter 11.

Millet, P. 2016. Conventional and innovative electrocatalysts for PEM water electrolysis. Pacific Rim Meeting on Electrochemical and Solid State Science (PRIME). In: Proceedings of the 16th Polymer Electrolyte Fuel Cell Symposium (PEFC16). The Electrochemical Society. 2016 October 2–7; Honolulu, Hawaii.

Millet, P., Durand, R., Pinéri, M., 1989. New solid polymer electrolyte composites for water electrolysis. J. Appl. Electrochem. 19, 162–166.

Millet, P., Durand, R., Pinéri, M., 1990. Preparation of new solid polymer electrolyte composites for water electrolysis. Int. J. Hydrogen Energy 15, 245–253.

Millet, P., Alleau, T., Durand, R., 1993. Characterization of membrane-electrodes assemblies for solid polymer electrolyte water electrolysis. J. Appl. Electrochem. 23, 322–331.

Millet, P., Andolfatto, F., Durand, R., 1996. Design and performances of a solid polymer electrolyte water electrolyser. Int. J. Hydrogen Energy 21, 87–96.

Millet, P., Mbemba, N., Grigoriev, S.A., Fateev, V.N., Aukauloo, A., Etiévant, C., 2011. Electrochemical performances of PEM water electrolysis cells and perspectives. Int. J. Hydrogen Energy 36, 4134–4142.

Millet, P., de Guglielmo, F., Grigoriev, S.A., Porembskiy, V.I., 2012. Cell failure mechanisms in PEM water electrolyzers. Int. J. Hydrogen Energy 37, 17478–17487.

Renner, J., Ayers, K., Anderson, E., 2015. Proton exchange membrane electrolyzers stack and system design. In: Bessarabov, D., Wand, H., Li, H., Zhao, N. (Eds.), PEM Water Electrolysis for Hydrogen Production: Principles and Applications. CRC Press, Taylor & Francis, Boca Raton, pp. 157–178. Chapter 15.

Rozain, C., Millet, P., 2014. Electrochemical characterization of polymer electrolyte membrane water electrolysis cells. Electrochim. Acta 131, 160–167.

Rozain, C., Guillet, N., Mayousse, E., Millet, P., 2016a. Influence of iridium oxide loadings on the performance of PEM water electrolysis cells: Part II – Advanced anodic electrodes. Appl. Catal., B 182, 123–131.

Rozain, C., Mayousse, E., Guillet, N., Millet, P., 2016b. Influence of iridium oxide loadings on the performance of PEM water electrolysis cells: Part I – Pure IrO_2-based anodes. Appl. Catal., B 182, 153–160.

S++ Simulation Services Co., Murnau-Westried, Germany, http://www.splusplus.com/.

Stucki, S., Scherer, G.G., Schlagowski, S., Fischer, E., 1998. PEM water electrolysers: Evidence for membrane failure in 100kW demonstration plants. J. Appl. Electrochem. 28, 1041–1049.

Trasatti, S., Petri, O., 1992. Real surface area measurement in electrochemistry. J. Electroanal. Chem. 327, 353–376.

Tsotridis, G., Pilenga, A., De Marco, G., Malkow, T., 2015. EU Harmonized test protocols for PEMFC MEA testing in single cell configuration for automotive applications. European Commission, JRC99115.

Verdin, B., 2018. Ph.D. Dissertation, Paris-Sud University, France.

Verdin, B., Bounoua, H., Moussabbir, A., Millet, P., 2017. Distribution of current lines in catalytic layers of PEM water electrolyzers. In: Proceedings of the 68th Annual Meeting of the International Society of Electrochemistry. 2017 August 27–September 1; Providence, Rhode Island, USA.

CHAPTER 4

Power-to-Gas

A great deal of progress has been made in the development of proton-exchange membrane (PEM) electrolyzers. New areas in technology development have been identified, as well as new applications, e.g., power-to-gas (P2G). One of the first descriptions of the P2G concept was presented by Sterner (2009) in his doctoral dissertation (focus on renewable power methane). Recently, several publications on P2G have appeared, e.g., Jentsch et al., 2014; Schiebahn et al., 2015; Chiuta et al., 2016; Estermann et al., 2016; Gotz et al., 2016; Scamman and Newborough 2016; Bailera et al., 2017a; 2017b; Mesfun et al., 2017; Meylan et al., 2017; Saric et al., 2017. Further analysis of PEM electrolyzer characteristics for the storage of renewable energy was very recently conducted by Kotowicz et al. (2017) and an algorithm for determining the thermodynamic and economic characteristics of a P2G installation is proposed.

According to Harvey et al. (2015), P2G is the term that describes the use of water electrolysis to convert surplus renewable electrical power generation into renewable hydrogen gas. Applications of P2G include hydrogen fuel for fuel cell electric vehicles, as a substitute for natural gas in pipelines, as renewable natural gas, or as a renewable hydrogen feedstock in the refining of conventional liquid fuels. By means of the existing gas grid, P2G systems can help balance the electricity grid and facilitate the utilization of renewable energy independent from the location of its origin (Simonis and Newborough, 2017). A detailed comparison of P2G with other energy storage technologies for large applications as well as possible energy services applications for P2G projects was given by Walker et al. (2016). The P2G concept requires large PEM electrolysis plants—and it is becoming a reality.

Based on Lehner et al. (2014), the P2G concept can be schematically represented as shown in Fig. 4.1.

There are, however, still technical and economic barriers that have to be overcome before P2G can become commercially successful

PEM Water Electrolysis. DOI: https://doi.org/10.1016/B978-0-08-102830-8.00004-7

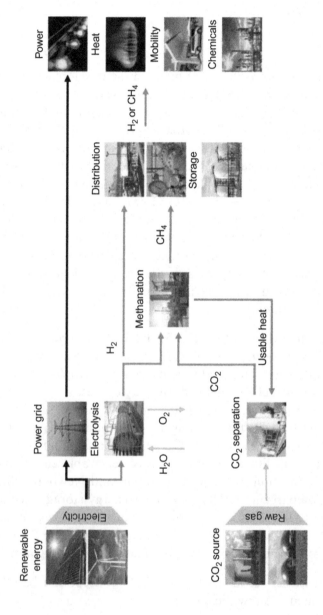

Figure 4.1 Schematic representation of the power-to-gas concept. (Inspired by Lehner et al., 2014; author's own compilation.)

Table 4.1 Summary of Currently Active and Proposed Power-to-Gas Projects Utilizing PEM Water Electrolysis Technology (Inspired by Bailera et al. (2017a), Updated With Authors' Own Compilation)

	Project	Installed Power (kW)	Methanation Principle	Application	Power Load	Ref. No/ Link
1	Reno, CO_2 recycling by reaction with renewably generated hydrogen	1	Chemical	Gas grid	Flexible load	1
2	Kjeller, IFE Project	1.75	n/a	Fuel cell	Flexible load	2
3	Neo Olvio of Xanthi, Systems Sunlight	4	n/a	Fuel cell	Flexible load	3
4	Boulder, Wind2H2 Project	5	n/a	Hydrogen IC engine and mobility	Flexible load	4
5	Reussenköge, Power Gap Filler Project	5	n/a	Gas turbine (biogas)	Flexible load	5
6	Porgrunn, NEXPEL Project	6	n/a	Flexible	Flexible load	6
7	Rozenburg, DNV GL	7	Chemical	Gas grid	Flexible load	7
8	Irvine, University of California P2G system	7 and 60	n/a	Gas grid	Flexible load	8
9	Stuttgart, Solarfuel	25	Chemical	Gas grid	Flexible load	9
10	Tahivilla, Hidrolica Project	40	n/a	Fuel cell	Flexible load	10
11	Bozcaada, Hydrogen Island	50	n/a	Fuel cells and mobility	Flexible load	11
12	Bella Coola, HARP Project	100	n/a	Fuel cell	Flexible load	12
13	Corsica Island, Myrte Project	110	n/a	Fuel cell	Flexible load	13
14	Herten, Stadt Herten & Evonic Industries	165	n/a	Mobility	Flexible load	14
15	Foulum, Electrochaea	250	Biological	Gas grid	Flexible load	15
16	Stuttgart, Solarfuel	250	Chemical	Gas grid	Flexible load	16
17	P2G BioMet Project	250	Biocatalytic	Gas grid	Flexible load	17
18	Cologne, RWE Power AG	300	Chemical	Gas grid	Flexible load	18
19	Frankfurt, Thuga & ITM Power	325	n/a	Gas grid	Flexible load	19
20	Nakhon Ratchasima, Lam Ta Khong Wind Turbine Generation Project; recently proposed	1000	n/a	Fuel cell	Flexible load	20
21	Zaragoza, MefCO2 Project	1000	n/a	Methanol synthesis	Flexible load	21

(Continued)

Table 4.1 (Continued)

	Project	Installed Power (kW)	Methanation Principle	Application	Power Load	Ref. No/ Link
22	Essen, MefCO₂ Project	1000	n/a	Methanol synthesis	Flexible load	22
23	Hobro, HyBalance Demonstration Project	1200	n/a	Flexible	Flexible load	23
24	Hamburg, E.ON AG	1500	n/a	Gas grid	Flexible load	24
25	Mainz, Energie Park	2000	n/a	Gas grid	Flexible load	25
26	Hannover, electrolyzer plant designs launched at Hannover MESSE 2017; recently proposed	2200	n/a	Petrochemical and gas grid	Flexible load	26
27	Brunsbüttel, Wind to Gas Südermarsch; recently proposed	2400	n/a	Gas grid	Flexible load	27
28	Linz, H2FUTURE Project; recently proposed	6000	n/a	Flexible	Flexible load	28
29	Wesseling, 10 MW Refinery Hydrogen Project; recently proposed	10000	n/a	Petrochemical	Base load	29

1. www.sciencedirect.com/science/article/pii/S1750583609001005
2. https://www.ife.no/en/ife/main_subjects_new/energy_environment/hydrogensystems
3. http://www.systems-sunlight.com/blog/on-line-energy-management-strategy-of-an-off-grid-hybrid-power-generation-system/
4. https://www.hydrogen.energy.gov/pdfs/review06/pd_7_kroposki.pdf
5. http://www.gp-joule.eu/news/newsdetails/schleswig-holsteins-first-power-to-gas-plant-goes-into-operation/
6. http://www.fch.europa.eu/project/next-generation-pem-electrolyser-sustainable-hydrogen-production
7. https://www.energyvalley.nl/downloads/dnv-gl-power2gas-demonstration-project-rozenburg
8. http://www.sciencedirect.com/science/article/pii/S1464285915301243
9. www.etogas.com
10. www.agenciaandaluzadelaenergia.es
11. https://fuelcellsworks.com/archives/2011/10/26/first-hydrogen-energy-production-on-a-turkish-island-has-started-on-bozcaada/
12. http://www.powertechlabs.com/temp/2016112384398/HARP_DataSheet_Feb_4_2011web.pdf
13. http://www.iphe.net/docs/Events/Seville_11-12/Workshop/Posters/IPHE%20workshop_MYRTE_poster.pdf
14. www.sotaventogalicia.com
15. www.electrochaea.com
16. www.etogas.com
17. http://www.cngservices.co.uk/images/BiomethaneDay/2013/UKBD2013_DH_P2G.pdf
18. http://www.rwe.com/web/cms/en/113648/rwe/press-news/press-release/?pmid=4008792
19. http://www.sciencedirect.com/science/article/pii/S1464285915300560.pdf
20. http://www.sciencedirect.com/science/article/pii/S1464285916302127
21. http://www.hydrogenics.com/about-the-company/news-updates/2015/01/26/hydrogenics-to-supply-1mw-pem-electrolyzer-to-european-consortium
22. http://www.sciencedirect.com/science/article/pii/S1464285915300237
23. http://hybalance.eu/latest-news-2/
24. http://www.hydrogenics.com/about-the-company/news-updates/2015/10/15/e.on-inaugurates-energy-storage-facility-using-hydrogenics-pem-technology
25. http://www.energiepark-mainz.de/en/technology/technology/
26. http://www.directorstalkinterviews.com/itm-power-plc-100mw-electrolyser-plant-designs-launched-hannover-zeus-comments/412718501
27. http://www.hydrogenics.com/2017/03/22/hydrogenics-awarded-2-4-mw-power-to-gas-plant-in-germany/
28. http://www.h2future-project.eu/technology
29. http://www.itm-power.com/news-item/10mw-refinery-hydrogen-project-with-shell

(Gotz et al., 2016). According to Lewandowska-Bernat and Desideri (2017), current challenges related to P2G technology include the lack of an appropriate policy framework as the grid system balancing technology and lack of detailed data for the technology, resulting in challenges with computer modeling.

Various technical requirements for electrolyzer plants, in the context of P2G applications, e.g., the discharge hydrogen pressure levels, were recently discussed by Bensmann et al. (2016). Specifically, it was mentioned that, for the direct injection of hydrogen into the natural gas grid, a hydrogen pressure of 5−85 bar is required.

In general, the attractive features of PEM water electrolysis technology for P2G application include the following: High dynamic operation, compact design, simple cold-start capability, ability to discharge hydrogen at high pressure, low degradation rates, and pure water handling—avoiding dealing with corrosive liquid electrolytes (Wolf, 2015).

Table 4.1 provides an updated summary of currently active and proposed P2G projects utilizing PEM water electrolysis technology.

REFERENCES

Bailera, M., Espatolero, S., Lisbona, P., Luis, M., Romeo, L.M., 2017a. Power to gas-electrochemical industry hybrid systems: a case study. Appl. Energy 202, 435−446.

Bailera, M., Lisbona, P., Romeo, L.M., Espatolero, S., 2017b. Power to gas projects review: Lab, pilot and demo plants for storing renewable energy and CO_2. Renewable Sustainable Energy Rev. 69, 292−312.

Bensmann, B., Hanke-Rauschenbach, R., Müller-Syring, G., Henel, M., Sundmacher, K., 2016. Optimal configuration and pressure levels of electrolyzer plants in context of power-to-gas applications. Appl. Energy 167, 107−124.

Chiuta, S., Engelbrecht, N., Human, G., Bessarabov, D.G., 2016. Technoeconomic assessment of power-to-methane and power-to-syngas business models for sustainable carbon dioxide utilization in coal-to-liquid facilities. J. CO2 Util 16, 399−411.

Estermann, T., Newborough, M., Sterner, M., 2016. Power-to-gas systems for absorbing excess solar power in electricity distribution networks. Int. J. Hydrogen Energy 41, 13950−13959.

Gotz, M., Lefebvre, J., Mors, F., McDaniel Koch, A., Graf, F., Bajohr, S., et al., 2016. Renewable power-to-gas: a technological and economic review. Renewable Energy 85, 1371−1390.

Harvey, R., Abouatallah, R., Cargnelli, J., 2015. Large-scale water electrolysis for power-to-gas. In: Bessarabov, D., Wang, H., Li, H., Zhao, N. (Eds.), PEM Electrolysis for Hydrogen Production: Principles and Applications, 313. CRC Press, Taylor & Francis Group, Boca Raton, p. 303.

Jentsch, M., Trost, T., Sterner, M., 2014. Optimal use of power-to-gas energy storage systems in an 85% renewable energy scenario. Energy Procedia 46, 254−261.

Kotowicz, J., Bartela, L., Wecel, D., Dubiel, K., 2017. Hydrogen generator characteristics for storage of renewably-generated energy. Energy 118, 156–171.

Lehner, M., Tichler, R., Steinmüller, H., Koppe, M., 2014. Power-to-Gas: Technology and Business Models. SpringerBriefs in Energy. Springer International Publishing, Cham.

Lewandowska-Bernat, A., Desideri, U., 2017. Opportunities of power-to-gas technology. Energy Procedia 105, 4569–4574.

Mesfun, S., Sanchez, D.L., Leduc, S., Wetterlund, E., Lundgren, J., Biberacher, M., et al., 2017. Power-to-gas and power-to-liquid for managing renewable electricity intermittency in the Alpine Region. Renewable Energy 107, 361–372.

Meylan, F.D., Piguet, F.P., Erkman, S., 2017. Power-to-gas through CO_2 methanation: Assessment of the carbon balance regarding EU directives. Energy Storage 11, 16–24.

Saric, M., Dijkstra, J.W., Haije, W.G., 2017. Economic perspectives of power-to-gas technologies in bio-methane production. J. CO2 Util. 20, 81–90.

Scamman, D., Newborough, M., 2016. Using surplus nuclear power for hydrogen mobility and power-to-gas in France. Int. J. Hydrogen Energy 6, 10080–10089.

Schiebahn, S., Grube, T., Robinus, M., Tietze, V., Kumar, B., Stolten, D., 2015. Power to gas: Technological overview, systems analysis and economic assessment for a case study in Germany. Int. J. Hydrogen Energy 40, 4285–4294.

Simonis, B., Newborough, M., 2017. Sizing and operating power-to-gas systems to absorb excess renewable electricity. Int. J. Hydrogen Energy 42, 21635–21647.

Sterner, M., 2009. Bioenergy and renewable power methane in integrated 100% renewable energy systems. Dissertation. Kassel University Press, Kassel.

Walker, S.B., Mukherjee, U., Fowler, M., Elkamel, A., 2016. Benchmarking and selection of Power-to-Gas utilizing electrolytic hydrogen as an energy storage alternative. Int. J. Hydrogen Energy 41, 7717–7731.

Wolf, E., 2015. Large-scale hydrogen energy storage. Chapter 9. In: Moseley,, P.T., Garche, J. (Eds.), Electrochemical Energy Storage for Renewable Sources and Grid Balancing. Elsevier, Amsterdam, pp. 129–142.

CHAPTER 5

Selected Properties of Hydrogen

5.1 DIFFUSION COEFFICIENT OF H_2 IN WATER

Jähne et al. (1987) fitted the Arrhenius equation (Eq. 5.1) to experimental data obtained for the diffusion of H_2 in water. Appropriate Arrhenius constants $(D_0 = 3.338E-2 \ cm^2.s^{-1}$ and $E_D = 16.06 \ kJ.mol^{-1})$ were identified for experimental data in the temperature range 5–35°C.

$$D = D_0 \exp\left(-\frac{E_D}{RT}\right) \qquad (5.1)$$

For additional reading refer to Himmelblau (1964) (Table 5.1).

5.2 SOLUBILITY OF H_2 IN WATER (TABLE 5.2)

The solubility of H_2 in water is represented by correlations derived by various authors, using different units.

- Bunsen coefficient, β ($cm^3 \ H_2$ (STP).$cm^{-3} \ H_2O$):

Crozier and Yamamoto (1974) fitted an expression (Eq. 5.2) to their own experimental data to determine the Bunsen coefficient as a function of temperature and salinity (parts per thousand). This expression is valid for temperatures of −2 to 30°C at 1 atm.

$$\ln \beta = -39.9611 + 53.9381 \left(\frac{100}{T}\right) + 16.3135 \ln\left(\frac{T}{100}\right)$$
$$+ S\left[-0.036249 + 0.017565\left(\frac{T}{100}\right) - 0.0023010\left(\frac{T}{100}\right)^2\right] \qquad (5.2)$$

Gordon et al. (1977) determined that their experimental data on the Bunsen coefficient is on average 0.5% higher than that of Crozier and Yamamoto (1974). Combining the data sets, the expression reported

PEM Water Electrolysis. DOI: https://doi.org/10.1016/B978-0-08-102830-8.00005-9

Table 5.1 Diffusion Coefficients of H_2 in Water as a Function of Temperature

$$D^*10E5 \ (cm^2.s^{-1})$$

Temperatures (°C)	Values	References	Temperatures (°C)	Values	References
0	1.96	Tammann and Jessen (1929)	23	3.85	Aikazyan and Fedorova (1952)
5	3.22	Jähne et al. (1987)	24.5	4.90	Houghton et al. (1962)
10	2.80	Tammann and Jessen (1929)	25	5.13	Jähne et al. (1987)
	3.64	Jähne et al. (1987)		3.37	Ipat'ev et al. (1933)
	4.34	Hufner (1898)		3.49	Ipat'ev et al. (1933)
13	6.72	Exner (1875)		7.07	Cullen and Davidson (1957)
14	3.41	Hagenbach (1898)	25.5	4.08	Tammann and Jessen (1929)
15	4.09	Jähne et al. (1987)	30	4.49	Tammann and Jessen (1929)
	2.49	Ipat'ev et al. (1933)		5.70	Jähne et al. (1987)
16	4.73	Hufner (1896)	35	6.32	Jähne et al. (1987)
17	5.15	Cullen and Davidson (1957)		4.22	Ipat'ev et al. (1933)
17.5	3.40	Tammann and Jessen (1929)	45	5.69	Ipat'ev et al. (1933)
20	4.59	Jähne et al. (1987)	100	23.2	Ipat'ev et al. (1933)
21	5.15	Hufner (1898)			

Source: Adapted from Himmelblau, D.M., 1964. Diffusion of gases in liquids. Chem. Rev. 64 (5), 527–550.

Table 5.2 Solubility of H_2 in Water as a Function of Temperature (1 atm Unless Specified Otherwise[a])

Temperatures (°C)	Units as Published	Values	$S^a10E4 \ cm^3 \ H_{2(STP)}.cm^{-3} \ H_2O.cmHg^{-1}$	References
15	Bunsen coefficient ($cm^3 \ H_{2(STP)}.cm^{-3} \ H_2O$)	1.873E−02	2.46	Crozier and Yamamoto (1974)
	Bunsen coefficient ($cm^3 \ H_{2(STP)}.cm^{-3} \ H_2O$)	1.877E−02	2.47	Gordon et al. (1977)
	Mole fraction	1.510E−05	2.47	Young (1981)
	atm.mole fraction^{-1}	6.675E+04	2.45	Himmelblau (1960)
	Henry constant (L.atm. mol^{-1})	1.215E+03	2.46	Kolev (2007)[a]
	$cm^3 \ H_{2(STP)}.cm^{-3} \ H_2O$	1.883E−02	2.48	Dean (1999)
20	Bunsen coefficient ($cm^3 \ H_{2(STP)}.cm^{-3} \ H_2O$)	1.802E−02	2.37	Crozier and Yamamoto (1974)
	Bunsen coefficient ($cm^3 \ H_{2(STP)}.cm^{-3} \ H_2O$)	1.808E−02	2.38	Gordon et al. (1977)

(Continued)

Table 5.2 (Continued)

Temperatures (°C)	Units as Published	Values	$S^a 10E4\ cm^3\ H_{2(STP)}.cm^{-3}\ H_2O.cmHg^{-1}$	References
	Mole fraction	1.455E−05	2.38	Young (1981)
	atm.mole fraction^{-1}	6.926E + 04	2.36	Himmelblau (1960)
	Henry constant (L.atm. mol^{-1})	1.258E + 03	2.38	Kolev (2007)[a]
	$cm^3\ H_{2(STP)}.cm^{-3}\ H2O$	1.819E−02	2.39	Dean (1999)
25	Bunsen coefficient ($cm^3\ H_{2(STP)}.cm^{-3}\ H_2O$)	1.744E−02	2.30	Crozier and Yamamoto (1974)
	Bunsen coefficient ($cm^3\ H_{2(STP)}.cm^{-3}\ H_2O$)	1.754E−02	2.31	Gordon et al. (1977)
	Mole fraction	1.412E−05	2.30	Young (1981)
	atm.mole fraction^{-1}	7.121E + 04	2.29	Himmelblau (1960)
	Henry constant (L.atm. mol^{-1})	1.301E + 03	2.30	Kolev (2007)[a]
	Concentration (mol.L^{-1})	7.800E−04	2.30	Hine and Weimar (1965)
	$cm^3\ H_{2(STP)}.cm^{-3}\ H_2O$	1.754E−02	2.31	Dean (1999)
30	Bunsen coefficient ($cm^3\ H_{2(STP)}.cm^{-3}\ H_2O$)	1.698E−02	2.23	Crozier and Yamamoto (1974)
	Bunsen coefficient ($cm^3\ H_{2(STP)}.cm^{-3}\ H_2O$)	1.713E−02	2.25	Gordon et al. (1977)
	Mole fraction	1.377E−05	2.24	Young (1981)
	atm.mole fraction^{-1}	7.306E + 04	2.23	Himmelblau (1960)
	Henry constant (L.atm. mol^{-1})	1.345E + 03	2.22	Kolev (2007)[a]
	$cm^3\ H_{2(STP)}.cm^{-3}\ H_2O$	1.699E−02	2.24	Dean (1999)
35	Mole fraction	1.350E−05	2.20	Young (1981)
	atm.mole fraction^{-1}	7.408E + 04	2.20	Himmelblau (1960)
	Henry constant (L.atm. mol^{-1})	1.389E + 03	2.15	Kolev (2007)[a]
	$cm^3\ H_{2(STP)}.cm^{-3}\ H_2O$	1.666E−02	2.19	Dean (1999)
40	Mole fraction	1.330E−05	2.16	Young (1981)
	atm.mole fraction^{-1}	7.490E + 04	2.17	Himmelblau (1960)
	Henry constant (L.atm. mol^{-1})	1.435E + 03	2.08	Kolev (2007)[a]
	$cm^3\ H_{2(STP)}.cm^{-3}\ H_2O$	1.644E−02	2.16	Dean (1999)

(Continued)

Table 5.2 (Continued)

Temperatures (°C)	Units as Published	Values	$S^a 10E4$ cm^3 H$_{2(STP)}$. cm^{-3} H$_2$O.cmHg^{-1}	References
45	Mole fraction	1.317E−05	2.14	Young (1981)
	atm.mole fraction^{-1}	7.539E + 04	2.15	Himmelblau (1960)
	Henry constant (L.atm. mol^{-1})	1.481E + 03	2.02	Kolev (2007)[a]
	cm^3 H2$_{(STP)}$.cm^{-3} H$_2$O	1.624E−02	2.14	Dean (1999)
50	Mole fraction	1.310E−05	2.14	Young (1981)
	atm.mole fraction^{-1}	7.558E + 04	2.16	Himmelblau (1960)
	Henry constant (L.atm. mol^{-1})	1.527E + 03	1.96	Kolev (2007)[a]
	cm^3 H$_{2(STP)}$.cm^{-3} H$_2$O	1.608E−02	2.12	Dean (1999)
60	Mole fraction	1.312E−05	2.11	Young (1981)
	atm.mole fraction^{-1}	7.511E + 04	2.14	Himmelblau (1960)
	Henry constant (L.atm. mol^{-1})	1.623E + 03	1.84	Kolev (2007)[a]
	cm^3 H$_{2(STP)}$.cm^{-3} H$_2$O	1.600E−02	2.11	Dean (1999)
80	Mole fraction	1.372E−05	2.18	Young (1981)
	atm.mole fraction^{-1}	7.171E + 04	2.22	Himmelblau (1960)
	Henry constant (L.atm.mol^{-1})	1.822E + 03	1.64	Kolev (2007)[a]
	cm^3 H$_{2(STP)}$.cm^{-3} H$_2$O	1.600E−02	2.11	Dean (1999)

[a] 1 bar.

by Crozier and Yamamoto (1974) was fitted with different numeric constants (Eq. 5.3).

$$\ln \beta = -49.641 + 67.460 \left(\frac{100}{T} \right) + 21.028 \ln \left(\frac{T}{100} \right)$$
$$+ S \left[-0.077314 + 0.046580 \left(\frac{T}{100} \right) - 0.0074291 \left(\frac{T}{100} \right)^2 \right]$$

$$(5.3)$$

- Mole fraction, X_{H_2}:
  Young (1981) fitted an equation (Eq. 5.4) to experimental data from 10 reliable literature sources for determining the mole fraction of H_2 dissolved in water at $0-80°C$ at 1 atm.

$$\ln X_{H_2} = -48.1611 + \frac{5528.45}{T} + 16.8893 \ln\left(\frac{T}{100}\right) \qquad (5.4)$$

- Henry's law constant, H (atm.mole fraction^{-1}):
  Himmelblau (1960) fitted a second-order expression (Eq. 5.5) to reliable literature values to determine the Henry's law constant of H_2 dissolved in water.

$$-0.1233\left(\log \frac{H}{10000}\right)^2 - 0.1366\left(\frac{1000}{T}\right)^2$$
$$+0.02155\left(\log \frac{H}{10000}\right)\left(\frac{1000}{T}\right) - 0.2368\left(\log \frac{H}{10000}\right) \qquad (5.5)$$
$$+0.8249\left(\frac{1000}{T}\right) - 1 = 0.$$

- Henry's law constant, k_H (L.atm.mol^{-1}):

  Kolev (2007) fitted a polynomial expression (Eq. 5.6) to the Henry's law constant for H_2 dissolved in water. This expression is valid for $0-100°C$ and 1 bar.

$$k_H = 1 \times 10^{-5} Pa \times (-1.543218 \times 10^6 + 1.3585 \times 10^4(T)$$
$$+ \left(3.78843 \times 10^1\right)^2(T)^2 + \left(3.51564 \times 10^{-2}\right)^3(T)^3). \qquad (5.6)$$

5.3 SOLUBILITY OF H_2 IN WATER WHEN TOTAL GAS PRESSURE INCLUDES WATER VAPOR EQUILIBRIUM

Gas partial pressure plus water vapor partial pressure equals 1 atm (Table 5.3).

Table 5.3 Solubility of H_2 in Water as a Function of Temperature for 1 atm Total Gas Pressure of H_2 and Water Vapor							
S*10E4 (cm³(STP).cm⁻³.cmHg⁻¹)							

Temperature (°C)							
0	5	10	15	20	25	30	35
2.806	2.663	2.540	2.433	2.336	2.234	2.143	2.068
40	45	50	60	70	80	90	100
2.005	1.939	1.856	1.691	1.456	1.121	6.483E−1	0

Source: *Adapted from Dean, J.A., 1999. Lange's Handbook of Chemistry. McGraw-Hill, New York*

5.4 ABSOLUTE HUMIDITY OF WATER IN H_2 AS A FUNCTION OF TEMPERATURE AND PRESSURE (TABLE 5.4)

Table 5.4 Absolute Humidity of Water in H_2 as a Function of Temperature and Pressure (kg H_2O.kg H_2^{-1})

		Pressure (atm)			
		1	10	50	100
Temperature (°C)	0	5.42E−02	5.39E−03	1.08E−03	5.38E−04
	10	1.10E−01	1.08E−02	2.17E−03	1.08E−03
	20	2.11E−01	2.07E−02	4.12E−03	2.06E−03
	30	3.90E−01	3.76E−02	7.49E−03	3.74E−03
	40	7.00E−01	6.56E−02	1.30E−02	6.51E−03
	50	1.24E + 00	1.10E−01	2.18E−02	1.09E−02
	60	2.18E + 00	1.79E−01	3.53E−02	1.76E−02
	70	3.94E + 00	2.84E−01	5.53E−02	2.76E−02
	80	7.73E + 00	4.39E−01	8.44E−02	4.20E−02
	90	1.94E + 01	6.65E−01	1.25E−01	6.22E−02
	100		9.94E−01	1.82E−01	9.02E−02

Sources: *References for H_2-H_2O virial coefficients are Harvey, A.H., Lemmon, E.W., 2004. Correlation for the second virial coefficient of water. J. Phys. Chem. Ref. Data 33 (1), 369–376; Hodges, M.P., Wheatley, R.J., Schenter, G.K., Harvey, A.H., 2004. Intermolecular potential and second virial coefficient of the water-hydrogen complex. J. Chem. Phys. 120 (2), 710–720.*

$$RT\left[\ln\left(\frac{x_w P}{e_s}\right)\right] = -RTk_H(1-x_w)P + (v_w - B_{ww})(P - e_s)$$

$$+ (B_{hh} - 2B_{hw} + B_{ww})(1-x_w)^2 P$$

$$\ln\left(\frac{k_H}{e_s}\right) = \frac{A}{T_R} + \frac{B\tau^{0.335}}{T_R} + \frac{C\exp(\tau)}{T_R^{0.41}}$$

$$T_R = \frac{T}{T_c}, \mathrm{Tc} = 647.096\mathrm{K}$$

$$\tau = 1 - T_R$$

$A = -4.732$

$B = 6.09$

$C = 6.061$.

5.5 SOLUBILITY OF H_2 IN WATER AS A FUNCTION OF TEMPERATURE AT ELEVATED PRESSURE (TABLE 5.5)

Table 5.5 Solubility of H_2 in Water as a Function of Temperature at Elevated Pressure

$S \times 10E4$ (cm^3(STP).cm^{-3}.cmHg^{-1})

Pressure (atm)	Temperature (°C)										
	0	10	20	30	40	50	60	70	80	90	100
25	2.820	2.561	2.369	2.251	2.190	2.164	2.167	2.201	2.274	2.388	2.531
50	2.804	2.545	2.353	2.235	2.174	2.150	2.162	2.194	2.265	2.372	2.498
75	2.799	2.541	2.349	2.232	2.171	2.145	2.154	2.189	2.256	2.351	2.472
100	2.789	2.531	2.342	2.222	2.163	2.137	2.145	2.181	2.247	2.343	2.466
150	2.759	2.502	2.312	2.195	2.136	2.112	2.121	2.158	2.228	2.325	2.436
200	2.728	2.475	2.285	2.169	2.110	2.089	2.101	2.140	2.205	2.297	2.410
300	2.653	2.414	2.236	2.129	2.071	2.058	2.066	2.101	2.167	2.260	2.355
400	2.584	2.358	2.191	2.088	2.036	2.018	2.030	2.066	2.126	2.209	2.305
500	2.528	2.311	2.148	2.050	2.001	1.986	1.998	2.033	2.089	2.167	2.262
600	2.479	2.266	2.110	2.017	1.969	1.952	1.961	1.997	2.051	2.139	2.226
700	2.433	2.227	2.077	1.984	1.937	1.921	1.933	1.966	2.020	2.096	2.189
800	2.380	2.184	2.040	1.953	1.908	1.891	1.902	1.936	1.992	2.067	2.151
900	2.323	2.140	2.006	1.923	1.880	1.866	1.876	1.908	1.962	2.033	2.112
1000	2.266	2.099	1.973	1.896	1.858	1.845	1.854	1.883	1.933	2.002	2.076

Source: Adapted from Wiebe, R., Gaddy, V.L., 1934. The solubility of hydrogen in water at 0, 50, 75 and 100° from 25 to 1000 atmospheres. J. Am. Chem. Soc. 56 (1), 76–79.

5.6 PERMEABILITY OF H₂ IN WATER

Data for H_2 permeability in water was calculated by Ito et al. (2011) for $0-80°C$ and 1 atm. The diffusion equation (Eq. 5.1) fitted by Wise and Houghton (1966) and the mole fraction solubility equation (Eq. 5.4) derived by Young (1981) were used (Table 5.6).

Table 5.6 Permeability of H₂ in Water as a Function of Temperature		
Temperatures (°C)	Values as Published P*10E16 (mol.cm^{-1}.s^{-1}.Pa^{-1})	Barrer
0	3.179	95.0
5	3.422	104.1
10	3.693	114.4
15	3.995	125.9
20	4.329	138.8
25	4.701	153.3
30	5.113	169.6
35	5.569	187.7
40	6.075	208.1
45	6.634	230.9
50	7.253	256.4
55	7.939	285.0
60	8.696	316.9
65	9.534	352.7
70	10.46	392.7
75	11.48	437.4
80	12.61	487.3

Sources: *Adapted from Ito, H., Maeda, T., Nakano, A., Takenaka, H., 2011. Properties of Nafion membranes under PEM water electrolysis conditions. Int. J. Hydrogen Energy, 36, 10527–10540; Wise, D.L., Houghton, G., 1966. The diffusion coefficient of ten slightly soluble gases in water at 10–60°C. Chem. Eng. Sci. 21, 999–1010; Young, C.L., 1981. Hydrogen and deuterium. In: IUPAC Solubility Data Series, Vol. 5/6. Oxford: Pergamon Press.*

5.7 VISCOSITY OF H$_2$ (TABLE 5.7)

Table 5.7 Dynamic Viscosity of H$_2$ as a Function of Temperature at 1 atm			
Temperatures (°C)	Values (μPa.s)	Temperatures (°C)	Values (μPa.s)
−200	3.337	60	9.617
−150	4.850	70	9.813
−100	6.139	80	10.007
−50	7.310	90	10.199
0	8.397	100	10.389
10	8.606	200	12.206
20	8.813	300	13.900
30	9.017	400	15.503
40	9.219	500	17.034
50	9.419	600	18.506

Source: From Lemmon, E.W., McLinden, M.O., Friend, D.G. Thermophysical properties of fluid systems. In: Linstrom P.J., Mallard, W.G. (Eds.), NIST Chemistry WebBook, NIST Standard Reference Database, Number 69. National Institute of Standards and Technology, Gaithersburg, MD. Available from: <http://webbook.nist.gov/chemistry/fluid/>.

5.8 THERMAL CONDUCTIVITY OF H$_2$ (TABLE 5.8)

Table 5.8 Thermal Conductivity of H$_2$ as a Function of Temperature at 1 atm			
Temperatures (°C)	Values (W.m^{-1}.K^{-1})	Temperatures (°C)	Values (W.m^{-1}.K^{-1})
−200	0.053	60	0.202
−150	0.084	70	0.207
−100	0.116	80	0.212
−50	0.146	90	0.217
0	0.173	100	0.221
10	0.178	200	0.268
20	0.182	300	0.315
30	0.187	400	0.363
40	0.192	500	0.412
50	0.197	600	0.461

Source: From Lemmon, E.W., McLinden, M.O., Friend, D.G. Thermophysical properties of fluid systems. In: Linstrom P.J., Mallard, W.G. (Eds.), NIST Chemistry WebBook, NIST Standard Reference Database, Number 69. National Institute of Standards and Technology, Gaithersburg, MD. Available from: <http://webbook.nist.gov/chemistry/fluid/>.

5.9 PHYSICAL AND THERMODYNAMIC PROPERTIES OF H₂ (TABLE 5.9)

Table 5.9 Physical and Thermodynamic Properties of H_2			
Constants	Units	Values	References
Molecular weight, MW	g.mol^{-1}	2.016	Haynes and Lide (2011)
Density, rho (25°C and 1 atm)	g.cm^{-3}	0.082	Haynes and Lide (2011)
Melting point, T_m	K	13.99	Haynes and Lide (2011)
Boiling point, T_b (1 atm)	K	20.37	Lemmon et al.
Critical temperature, T_c	K	33.15	Lemmon et al.
Critical pressure, T_p	MPa	1.2964	Lemmon et al.
Heat capacity at constant P, Cp (25°C and 1 atm)	J.mol^{-1}.K^{-1}	28.84	Lemmon et al.
Heat capacity at constant V, Cv (25°C and 1 atm)	J.mol^{-1}.K^{-1}	20.52	Lemmon et al.
Ratio of specific heats, Cp/Cv (25°C and 1 atm)	unitless	1.405	Lemmon et al.
Joule-Thompson coefficient (25°C and 1 atm)	K.MPa^{-1}	−0.299	Lemmon et al.
Thermal conductivity, k (25°C and 1 atm)	W.m^{-1}.K^{-1}	0.185	Lemmon et al.
Relative thermal conductivity to air (25°C and 1 atm)	unitless	7.139	Lemmon et al.
Diffusion coefficient in air	cm^2.s^{-1}	0.61	Davis et al. (2013)
Higher heating value, HHV	MJ.kg^{-1}	141,6	Davis et al. (2013)
Lower heating value, LHV	MJ.kg^{-1}	119,6	Davis et al. (2013)
Auto-ignition temperature	K	858	Winter (1990)
Burning velocity in air (NTP)	m.s^{-1}	2.65−3.25	Winter (1990)
Minimum ignition energy in air	mJ	0.02	Winter (1990)
Flammability limits in air	%	4.0−75.0	Winter (1990)
Adiabatic flame temperature in air	K	2169	Fristrom (1995)
Adiabatic flame temperature in O_2	K	3000	Fristrom (1995)

5.10 ENERGY DENSITY OF H₂ COMPARED TO OTHER COMMON FUELS (TABLE 5.10)

Table 5.10 Thermo-Physical Properties of Common Fuels

Fuel	Density (kg.m^{-3})	LHV (MJ.kg^{-1})	LHV (MJ.L^{-1})	References
H$_2$ (1 atm)	0.082	119.60	0.00981	Haynes and Lide (2011), Davis et al. (2013)
H$_2$ (350 bar)	23.30	119.60	2.79	Davis et al. (2013), Hirscher (2010)
H$_2$ (700 bar)	39.30	119.60	4.70	Davis et al. (2013), Hirscher (2010)
Liquefied H$_2$ (1 atm)	70.80	119.6	8.47	Züttel (2004)
Natural gas (1 atm)	0.734	46.89	0.03442	NIST (2013)
Liquefied natural gas (1 atm)	456	49.41	22.53	NIST (2013), NETL (2005)
Methanol	792	19.99	15.83	John and Hansen (2014)
Ethanol	785	26.87	21.09	John and Hansen (2014)
Dimethyl ether (1 atm, $-25°C$)	661	28.62	18.92	John and Hansen (2014)
Gasoline (C$_7$H$_{16}$)	737	43.47	32.04	John and Hansen (2014)
Diesel (C$_{14}$H$_{30}$)	856	41.66	35.66	John and Hansen (2014)
Biodiesel (B100)	874	38.09	33.32	Alleman et al. (2016)

5.11 EMISSIONS OF COMMON COMBUSTION FUELS (TABLE 5.11)

Table 5.11 Emissions of Common Combustion Fuels (kg/kg Fuel Burned)				
	Fuel			
	H_2	C	CH_4	C_8H_{17}
CO_2	0	1.893	2.75	3.09
SO_2	0	0.012	0.03	0.01
N_xO	0.016	0.008	0.0075	0.0115
Dust and unburned matter	0	0.10	0	0.85
H_2O	7.00	0.633	2.154	1.254
$Pb(C_2H_5)_4$	0	0	0	0.001

Sources: *Adapted from Arienti, R., Nicoletti, G., 1990. La definizione di un indice entropico ambientale di alcuni combustibili. In: Proceedings of the 45th Italian Conference ATI, Cagliari, Italy; Nicoletti, G., 1995. The hydrogen option for energy: a review of technical, environmental and economic aspects. Int. J. Hydrogen Energy 20 (10), 759–765.*

5.12 INTERNAL ENERGY, ENTHALPY AND ENTROPY OF H_2 AT VARIOUS TEMPERATURE AND PRESSURE CONDITIONS (TABLE 5.12)

Table 5.12 Internal Energy, Enthalpy and Entropy of H_2 at Various Temperature and Pressure Conditions			
Temperatures (°C)	Internal Energy (kJ.mol^{-1})	Enthalpy (kJ.mol^{-1})	Entropy (J.mol^{-1}.K^{-1})
P = 1 atm			
−200	1.4333	2.0403	72.180
−150	2.1452	3.1695	83.884
−100	2.9927	4.4333	92.474
−50	3.9369	5.7935	99.368
0	4.9351	7.2076	105.08
10	5.1386	7.4943	106.11
20	5.3431	7.7819	107.11
30	5.5483	8.0703	108.08
40	5.7542	8.3594	109.02
50	5.9607	8.6490	109.93
60	6.1676	8.9391	110.81
70	6.3750	9.2296	111.67
80	6.5827	9.5205	112.51
90	6.7906	9.8116	113.32
100	6.9988	10.103	114.11
200	9.0880	13.024	121.05
300	11.184	15.951	126.66
400	13.288	18.886	131.38
500	15.407	21.837	135.46
600	17.546	24.807	139.08
P = 10 atm			
−200	1.3948	1.9908	52.512
−150	2.1255	3.1532	64.581
−100	2.9802	4.4295	73.258
−50	3.9282	5.7962	80.184
0	4.9289	7.2142	85.915
10	5.1328	7.5015	86.948
20	5.3376	7.7896	87.949
30	5.5431	8.0785	88.918
40	5.7493	8.3681	89.857
50	5.9561	8.6581	90.769
60	6.1633	8.9486	91.654
70	6.3709	9.2395	92.515
80	6.5788	9.5307	93.351

(Continued)

Table 5.12 (Continued)			
Temperatures (°C)	Internal Energy (kJ.mol^{-1})	Enthalpy (kJ.mol^{-1})	Entropy (J.mol^{-1}.K^{-1})
90	6.7870	9.8222	94.165
100	6.9955	10.114	94.957
200	9.0862	13.037	101.90
300	11.183	15.965	107.51
400	13.288	18.902	112.23
500	15.407	21.852	116.32
600	17.547	24.824	119.93
P = 100 atm			
−200	1.0472	1.6528	29.561
−150	1.9486	3.0484	44.173
−100	2.8669	4.4203	53.514
−50	3.8492	5.8390	60.707
0	4.8716	7.2894	66.571
10	5.0790	7.5818	67.622
20	5.2870	7.8746	68.638
30	5.4956	8.1679	69.622
40	5.7046	8.4615	70.575
50	5.9140	8.7553	71.498
60	6.1237	9.0493	72.394
70	6.3337	9.3435	73.265
80	6.5438	9.6378	74.110
90	6.7541	9.9322	74.932
100	6.9646	10.227	75.732
200	9.0708	13.170	82.719
300	11.178	16.110	88.357
400	13.290	19.054	93.091
500	15.414	22.009	97.184
600	17.558	24.983	100.80

Reference internal energy at 273,16 K for a saturated liquid.
Reference entropy at 273,16 K for a saturated liquid.
Source: *Lemmon, E.W., McLinden, M.O., Friend, D.G. Thermophysical properties of fluid systems. In: Linstrom, P.J., Mallard, W.G. (Eds.), NIST Chemistry WebBook, NIST Standard Reference Database, Number 69. National Institute of Standards and Technology, Gaithersburg, MD. Available from: <http:// webbook.nist.gov/chemistry/fluid/ > Lemmon et al.*

5.13 PERMEABILITY OF H$_2$ IN COMMON POLYMERS (TABLE 5.13)

Table 5.13 Permeability of H$_2$ in Common Polymers as a Function of Temperature					
Materials	Temperatures (°C)	Units as Published	Values	Barrer	References
PP, poly(propylene), 0.907 g.cm^{-3}, 50% crystallinity	20	cm^3$_{(STP)}$.cm. cm^{-2}.s^{-1}. Pa^{-1}	3.10E−12	41.33	(Miyake et al., 1983)
LDPE, poly(ethylene), 0.914 g.cm^{-3}	25	cm^3$_{(STP)}$.cm. cm^{-2}.s^{-1}. Pa^{-1}	7.40E−13	9.87	Miyake et al. (1983)
HDPE, poly(ethylene)	25	cm^3$_{(STP)}$.mm. m^{-2}.day^{-1}. atm^{-1}	154	2.35	Massey (2003)
PVC, poly(vinyl chloride), unplasticized	25	cm^3$_{(STP)}$.cm. cm^{-2}.s^{-1}. Pa^{-1}	1.30E−13	1.73	Miyake et al. (1983)
PVDF, poly(vinylidene fluoride)	23	cm^3$_{(STP)}$.mm. m^{-2}.day^{-1}. atm^{-1}	21.3	0.32	Massey (2003)
PTFE, poly (tetrafluoroethylene)	25	cm^3$_{(STP)}$.cm. cm^{-2}.s^{-1}. cmHg^{-1}	9.80E−10	9.80	Pasternak et al. (1970)
	25	cm^3$_{(STP)}$.cm. cm^{-2}.s^{-1}. Pa^{-1}	7.25E−13	9.67	Miyake et al. (1983)
Synthetic rubber, poly (butadiene)	25	cm^3$_{(STP)}$.cm. cm^{-2}.s^{-1}. atm^{-1}	3.20E−07	42.11	van Amerongen (1950)
	50	cm^3$_{(STP)}$.cm. cm^{-2}.s^{-1}. atm^{-1}	7.70E−07	101.32	van Amerongen (1950)
Natural rubber, poly (isoprene), amorphous	25	cm^3$_{(STP)}$.cm. cm^{-2}.s^{-1}. atm^{-1}	3.74E−07	49.21	van Amerongen (1950)
	50	cm^3$_{(STP)}$.cm. cm^{-2}.s^{-1}. atm^{-1}	9.08E−07	119.47	van Amerongen (1950)
Methyl rubber, poly (dimethylbutadiene)	25	cm^3$_{(STP)}$.cm. cm^{-2}.s^{-1}. Pa^{-1}	1.28E−12	17.07	Miyake et al. (1983)
Butyl rubber	25	cm^3$_{(STP)}$.cm. cm^{-2}.s^{-1}. atm^{-1}	5.50E−08	7.24	van Amerongen (1950)
	50	cm^3$_{(STP)}$.cm. cm^{-2}.s^{-1}. atm^{-1}	1.72E−07	22.63	van Amerongen (1950)

(Continued)

Table 5.13 (Continued)

Materials	Temperatures (°C)	Units as Published	Values	Barrer	References
Perbutan 18, poly (butadiene−co−acrylonitrile), 80/20	25	$cm^3_{(STP)}.cm.$ $cm^{-2}.s^{-1}.$ atm^{-1}	1.92E−07	25.26	van Amerongen (1950)
	50	$cm^3_{(STP)}.cm.$ $cm^{-2}.s^{-1}.$ atm^{-1}	4.91E−07	64.61	van Amerongen (1950)
Perbutan German, poly (butadiene−co−acrylonitrile), 73/27	25	$cm^3_{(STP)}.cm.$ $cm^{-2}.s^{-1}.$ atm^{-1}	1.21E−07	15.92	van Amerongen (1950)
	50	$cm^3_{(STP)}.cm.$ $cm^{-2}.s^{-1}.$ atm^{-1}	3.37E−07	44.34	van Amerongen (1950)
Hycar OR 25, poly (butadiene−co−acrylonitrile), 68/32	25	$cm^3_{(STP)}.cm.$ $cm^{-2}.s^{-1}.$ atm^{-1}	8.97E−08	11.80	van Amerongen (1950)
	50	$cm^3_{(STP)}.cm.$ $cm^{-2}.s^{-1}.$ atm^{-1}	2.63E−07	34.61	van Amerongen (1950)
Hycar OR 15, poly (butadiene−co−acrylonitrile), 61/39	25	$cm^3_{(STP)}.cm.$ $cm^{-2}.s^{-1}.$ atm^{-1}	5.42E−08	7.13	van Amerongen (1950)
	50	$cm^3_{(STP)}.cm.$ $cm^{-2}.s^{-1}.$ atm^{-1}	1.70E−07	22.37	van Amerongen (1950)
Neoprene G	25	$ml.cm.cm^{-2}.$ min^{-1}	5.40E−06	11.84	Sager (1940)
	25	$cm^3_{(STP)}.cm.$ $cm^{-2}.s^{-1}.$ Pa^{-1}	1.02E−12	13.60	van Amerongen (1946); van Amerongen (1964)
Polycarbonate	25	$m^3_{(STP)}.m.$ $m^{-2}.s^{-1}.Pa^{-1}$	9.00E−17	12.00	Norton (1963)
Kynar, poly(vinyl fluoride)	35	$cm^3_{(STP)}.cm.$ $cm^{-2}.s^{-1}.$ Pa^{-1}	4.04E−14	0.54	Miyake et al. (1983)
Poly(dimethyl siloxane)	35	$cm^3_{(STP)}.cm.$ $cm^{-2}.s^{-1}.$ Pa^{-1}	7.05E−11	939.92	Miyake et al. (1983)
Neoprene	35	$cm^3_{(STP)}.cm.$ $cm^{-2}.s^{-1}.$ $cmHg^{-1}$	2.76E−09	27.60	Fitch et al. (1993)
Hypalon40 (chlorosulfonated polyethylene)	35	$cm^3_{(STP)}.cm.$ $cm^{-2}.s^{-1}.$ $cmHg^{-1}$	1.10E−09	11.00	Fitch et al. (1993)
Hypalon45 (chlorosulfonated polyethylene)	35	$cm^3_{(STP)}.cm.$ $cm^{-2}.s^{-1}.$ $cmHg^{-1}$	1.56E−09	15.60	Fitch et al. (1993)

(*Continued*)

Table 5.13 (Continued)					
Materials	Temperatures (°C)	Units as Published	Values	Barrer	References
VitonE60 (fluoroelastomer)	35	cm$^3_{(STP)}$.cm. cm^{-2}.s^{-1}. cmHg^{-1}	1.06E−09	10.60	Fitch et al. (1993)
VitonGF (fluoroelastomer)	35	cm$^3_{(STP)}$.cm. cm^{-2}.s^{-1}. cmHg^{-1}	2.18E−09	21.80	Fitch et al. (1993)

5.14 VAN DER WAALS EQUATION OF STATE FOR REAL GASES

$$\left(P + \frac{n^2 a}{V^2}\right)(V - nb) = nRT$$

$$a = \frac{27 R^2 T_c^2}{64 P_c}$$

$$b = \frac{RT_c}{8 P_c}$$

P = pressure (bar)
V = volume (L)
n = amount of gas (mol)
R = gas constant (0.08314 bar.L.mol^{-1}.K^{-1})
For hydrogen:
   a = 0.2452 bar.L^2.mol^{-2}
   b = 0.0265 L.mol^{-1}

REFERENCES

Aikazyan, E.A., Fedorova, A.I., 1952. Electrochemical determination of the diffusion coefficient of hydrogen. Dokl. Akad. Nauk. S.S.S.R. 86, 1137–1140.

Alleman, T.L., McCormick, R.L., Christensen, E.D., Fioroni, G., Moriarty, K., Yanowitz, J., 2016. Biodiesel Handling and Use Guide, fifth ed. Report DOE/GO-102016-4875. National Renewable Energy Laboratory, Golden, CO. Available from: <http://biodiesel.org/docs/using-hotline/nrel-handling-and-use.pdf?sfvrsn=4>.

van Amerongen, G.J., 1946. Permeability of different rubbers to gases and its relation to diffusivity and solubility. J. Appl. Phys. 17, 972–985.

van Amerongen, G.J., 1950. Influence of structure of elastomers on their permeability to gases. J. Polym. Sci. 5, 307–332.

van Amerongen, G.J., 1964. Diffusion in elastomers. Rubber Chem. Technol. 37 (5), 1065–1152.

Arienti, R., Nicoletti, G., 1990. La definizione di un indice entropico ambientale di alcuni combustibili. In proceedings of the 45th Italian Conference ATI, Cagliari, Italy.

Crozier, T.E., Yamamoto, S., 1974. Solubility of hydrogen in water, seawater and NaCl solutions. J. Chem. Eng. Data 19 (3), 242–244.

Cullen, E.J., Davidson, J.F., 1957. The determination of diffusion coefficients for sparingly soluble gases in liquids. Trans. Inst. Chem. Eng. 35, 51–60.

Davis, S.C., Diegel, S.W., Boundy, R.G., 2013. Transportation energy data book. U.S. Department of Energy. Available from: <http://cta.ornl.gov/data>.

Dean, J.A., 1999. Lange's Handbook of Chemistry. McGraw-Hill, New York.

Exner, F., 1875. Über den Durchgang der Gase durch Flüssigkeitslamellen. Pogg. Ann. 155, 321–336.

Fitch, M.W., Koros, W.J., Nolen, R.L., Carnes, J.R., 1993. Permeation of several gases through elastomers, with emphasis on the deuterium/hydrogen pair. J. Appl. Polym. Sci. 47, 1033–1046.

Fristrom, R.M., 1995. Flame Structures and Processes. Oxford University Press, New York.

Gordon, L.I., Cohen, Y., Standley, D.R., 1977. The solubility of molecular hydrogen in seawater. Deep Sea Res. 24 (10), 937–941.

Hagenbach, A., 1898. Über Diffusion von Gasen durch wasserhaltige Gelatine. Wied. Ann. 65, 673–706.

Haynes, W.M., Lide, D.R., 2011. Handbook of Chemistry and Physics. CRC Press, Boca Raton, FL.

Himmelblau, D.M., 1960. Solubilities of inert gases in water. J. Chem. Eng. Data 5 (1), 10–15.

Himmelblau, D.M., 1964. Diffusion of gases in liquids. Chem. Rev. 64 (5), 527–550.

Hine, J., Weimar, R.D., 1965. Carbon basicity. J. Am. Chem. Soc. 87 (15), 3387–3396.

Hirscher, M., 2010. Handbook of Hydrogen Storage: New Materials for Future Energy Storage. Wiley-VCH, Weinheim, Germany.

Houghton, G., Ritchie, P.D., Thomson, J.A., 1962. The rate of solution of small stationary bubbles and the diffusion coefficients of gases in liquids. Chem. Eng. Sci. 17, 221–227.

Hufner, G., 1896. Ueber die Bestimmung der Diffusionscoefficienten einiger Gase für Wasser. Ann. Phys. 296, 134–168.

Hufner, G., 1898. Über die Diffusion von Gasen durch Wasser und durch Agargallerte. Z. Physik. Chem. 27, 227–249.

Ipat'ev, V.V., Teodorovich, V.P., Drushina-Artemovich, S.I., 1933. Diffusion of gases in liquids under pressure. Z. Anorg. Allgem. Chem 216, 66–74.

Jähne, B., Heinz, G., Dietrich, W., 1987. Diffusion coefficients of gases in water. J. Geophys. Res. 92, 10767–10776.

John, B., Hansen, S., 2014. Fuels and fuel processing options for fuel cells. In: Proceedings of the 2nd International Fuel Cell Conference, Lucerne, Switzerland.

Kolev, N.I., 2007. Multiphase Flow Dynamics 3: Turbulence, Gas Absorption and Release, Diesel Fuel Properties. Springer-Verlag, Berlin.

Lemmon, E.W., McLinden, M.O., Friend, D.G., Thermophysical properties of fluid systems. In: Linstrom, P.J., Mallard, W.G. (Eds.), NIST Chemistry WebBook, NIST Standard Reference Database, Number 69. National Institute of Standards and Technology, Gaithersburg, MD. Available from: <http://webbook.nist.gov/chemistry/fluid/>.

Massey, L.K., 2003. Permeability Properties of Plastics and Elastomers, A Guide to Packaging and Barrier Materials. Plastics Design Library, Norwich, NY.

Miyake, H., Matsuyama, M., Ashida, K., Watanabe, K., 1983. Permeation, diffusion, and solution of hydrogen isotopes, methane, and inert gases in/through tetrafluoroethylene and polyethylene. J. Vac. Sci. Technol. A-I 3, 1447.

NETL, 2005. Liquefied Natural Gas: Understanding the Basic Facts. Report DOE/FE-0489. National Energy Technology Laboratory, Pittsburgh, PA. Available from: <https://energy.gov/sites/prod/files/2013/04/f0/LNG_primerupd.pdf>.

Nicoletti, G., 1995. The hydrogen option for energy: a review of technical, environmental and economic aspects. Int. J. Hydrogen Energy 20 (10), 759–765.

NIST, 2013. Report of the 98th National Conference on Weights and Measures, NIST Special Publication 1197. In: Butcher, T., Crown, L., Harshman, R., Sefcik, D. and Warfield, L. (Eds.), pp. 345–346. Available from: <https://www.nist.gov/sites/default/files/documents/2017/05/09/2013-annual-sp1171-final.pdf>.

Norton, F.J., 1963. Gas permeation through lexan polycarbonate resin. Appl. Polym. Sci. 7 (5), 1–649.

Pasternak, R.A., Christenson, M.V., Heller, J., 1970. Diffusion and permeation of oxygen, nitrogen, carbon dioxide, and nitrogen dioxide through polytetrafluoroethylene. Macromolecules 3, 366–371.

Sager, T.P., 1940. Permeability of elastic polymers to hydrogen. J. Res. Natl. Bur. Stand. 25 (3), 309–313.

Tammann, G., Jessen, V., 1929. Diffusion coefficients in water and their temperature dependence. Z. Anorg. Allgem. Chem. 179, 125–144.

Wiebe, R., Gaddy, V.L., 1934. The solubility of hydrogen in water at 0, 50, 75 and 100° from 25 to 1000 atmospheres. J. Am. Chem. Soc. 56 (1), 76–79.

Winter, C.J., 1990. Hydrogen in high speed air transportation. Int. J. Hydrogen Energy 15, 579–595.

Young, C.L., 1981. Hydrogen and deuterium, IUPAC Solubility Data Series, Vol. 5/6. Pergamon Press, Oxford.

Züttel, A., 2004. Hydrogen storage methods. Naturwissenschaften 91 (4), 157–172.

FURTHER READING

Fernandez-Prini, R., Alvarez, J.L., Harvey, A.H., 2003. Henry's constants and vapour-liquid distribution constants for gaseous solutes in H_2O and D_2O at high temperatures. J. Phys. Chem. Ref. Data 32, 903–916.

Huang, P.H., 2008. Humidity standard of compressed hydrogen for fuel cell technology. ECS Trans. 12 (1), 479–484.

IAPWS, 2007. International Association for the Properties of Water and Steam (IAPWS) Industrial Formulation 1997.

IAPWS, 2007. IAPWS Industrial Formulation for the Thermodynamic Properties of Water and Steam (IAPWS-IF97). International Association for the Properties of Water and Steam (IAPWS) Industrial Formulation, 1997.

Reid, R.C., Prausnitz, J.M., Poling, B.E., 1987. The Properties of Gases and Liquids, 4th ed. McGraw-Hill, New York.

INDEX

Printed in the United States
By Bookmasters